Mary Kirby, Elizabeth Kirby

Stories about birds of land and water

Mary Kirby, Elizabeth Kirby

Stories about birds of land and water

ISBN/EAN: 9783337143152

Printed in Europe, USA, Canada, Australia, Japan

Cover: Foto ©berggeist007 / pixelio.de

More available books at **www.hansebooks.com**

Stories about Birds

OF

Land and Water.

BY

M. AND E. KIRBY,

AUTHORS OF "CHAPTERS ON TREES," ETC.

WITH NUMEROUS ILLUSTRATIONS.

Second Edition.

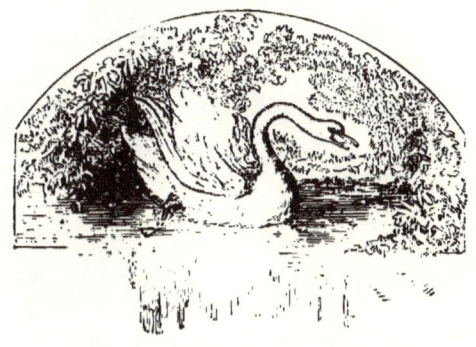

AMERICAN PUBLISHING CO.,
HARTFORD.

PREFACE.

THERE is not among the many beautiful creatures God has made one more to be admired than the bird.

Its graceful movements, its soft and elegant plumage, its gift of song—or if not of song, of a certain lively gaiety—its tender care for its young, its skill in preparing so pretty a home for their reception, its plaintive mourning when deprived of mate or young, its faith as, in search of a more genial clime, it flies across the mighty waters—all endear it to us as to a beautiful gift we ought to appreciate, and with whose happy and joyous life we should never wilfully interfere.

It is to you, dear children, we offer this little volume; read it,—and when you have done so, we think you will love the birds even better than before; and that when you want a few of their pretty eggs, you will not greedily them all, or tear down the cosy nest when the little pair are so happy ...at you will remember God is looking on—God, who cares for them as well as for you—who watches to see whether you obey His command, and spare the birds when you take their eggs, or whether you will hurt and destroy them.

Ah! How sad it would be if there were no birds!

Tip! tap! Listen—what's that? A robin on the window-sill. Open quickly and give him some crumbs!

M. AND E. K.

MELTON MOWBRAY.

CONTENTS.

	PAGE		PAGE
THE GOLDEN EAGLE	11	THE WHITE-THROAT	82
THE SEA EAGLE	16	THE LONG-TAILED TIT	84
THE TUFTED EAGLE	20	THE WAGTAIL	85
THE PEREGRINE FALCON	21	THE WREN	89
THE GOSHAWK	24	THE GOLDEN-CRESTED WREN	91
THE SECRETARY BIRD	26	THE GOLDEN ORIOLE	94
THE VULTURE	29	THE MOCKING-BIRD	94
THE CONDOR	32	THE SONG THRUSH	98
THE KITE	35	THE BLACKBIRD	100
THE BUZZARD	36	THE WATER OUSEL	102
THE SNAKE BUZZARD	40	THE LAUGHING THRUSH	105
THE SNOWY OWL	42	THE SHRIKE	106
THE SHORT-EARED OWL	47	THE STARLING	109
THE STONE OWL	48	THE SUPERB GLOSSY STARLING	112
THE GREAT HORNED OWL	50	THE CHOUGH	112
THE SWALLOW	53	THE RAVEN	115
THE MARTIN	58	THE CARRION CROW	119
THE KINGFISHER	61	THE ROOK	121
THE HOOPOE	64	THE JACKDAW	123
THE HUMMING-BIRD	67	THE MAGPIE	125
THE SUN-BIRD	71	THE NUTCRACKER	128
THE COMMON TREE-CREEPER	72	THE BIRD OF PARADISE	130
THE NIGHTINGALE	75	THE GOLDFINCH	132
THE ROBIN	76	THE CHAFFINCH	135
THE REDSTART	80	THE HOUSE SPARROW	137

CONTENTS.

	PAGE		PAGE
THE BROWN LINNET	141	THE QUAIL	204
THE LARK	144	THE CURLEW	206
THE CANARY	147	THE PLOVER	209
THE CUCKOO	150	THE LAPWING	212
THE PARROT	155	THE WOODCOCK	215
THE PARRAKEET	161	THE HERON	218
THE COCKATOO	164	THE STORK	225
THE BRUSH TURKEY	170	THE SPOON-BILL	232
THE CAPERCAILZIE	171	THE FLAMINGO	233
THE BLACK-COCK	175	THE PEACOCK	236
THE LYRE-BIRD	177	THE PELICAN	240
THE PHEASANT	180	THE CORMORANT	243
THE OSTRICH	185	THE SWAN	246
THE EMU	191	THE DUCK	250
THE BUSTARD	193	THE GOOSE	253
THE PRAIRIE HEN	197	THE STORMY PETREL	256
THE PTARMIGAN	201		

LIST OF ILLUSTRATIONS.

	PAGE
The Owl and its Prey	Frontispiece
The Golden Eagle	12
The Royal Eagle	13
The Imperial Eagle	15
The Sea Eagle	17
The Tufted Eagle	19
Peregrine Falcons	21
The Goshawk	24
The Sparrow-hawk	25
The Secretary Bird	27
The Vulture	29
The Tawny Goose Vulture	31
The Condor	33
The Common Kite	37
The Common Buzzard	39
The Snake Buzzard	41
The Snowy Owl	43
The Short-eared Owl	47
The Stone Owl	49
The Great Horned Owl	51
Swallows	55
The Martin	59
The Kingfisher	63
The Hoopoe	65
The Giant Humming-bird	67
The Sickle-billed Humming-bird	69
The Topaz-throated Humming-bird	70
Sun-birds	73
The Common Tree-creeper	74
The Nightingale	77
Robin Red-breast	79
The Redstart	80
White-throats	82
Long-tailed Tits	84
Wagtails	86
The White Wagtail	87
The Wren	89
Golden-crested Wrens	92
The Golden Oriole	93
The Mocking-bird	95
The Song Thrush	99
The Blackbird	101
The Water Ousel	103
The Laughing Thrush	105
The Magpie Shrike	107
The Rose Starling	110
The Superb Glossy Starling	111
The Chough	113
The Raven	117
The Carrion Crow	119
The Rook	121
The Jackdaw	123
The Magpie	127
The Nutcracker	129
The Red Bird of Paradise	131
Goldfinches	133
Chaffinches	136
The House Sparrow	137
Tree Sparrow and House Sparrow	139
The Brown Linnet	142
The Sky-lark	143
The Morn-lark	144
The Desert-lark	145
The Wild Canary	148
The Tame Canary	149
The Cuckoo	150
The Jay Cuckoo	151
The Golden Cuckoo	153

LIST OF ILLUSTRATIONS.

	PAGE
The Grey Parrot	156
The Amazon Parrot	157
The Waved Parrot	158
The Collared Parrot	159
The Ground Parrakeet	161
The Garuba Parrakeet	162
The Dappled Lorikeet	163
The Raven Cockatoo	165
The Nestor Cockatoo	166
The Helmet Cockatoo	167
The Casmalos Cockatoo	169
The Brush Turkey	170
The Capercailzie	173
The Black-cock	176
The Lyre-bird	179
The Silver Pheasant	181
The Black Pheasant	182
The Common Pheasant	183
Chinese Pheasants	184
The Ostrich	187
Ostrich Hunt	189
The American Ostrich	190
The Emu	192
Bustards	194
The Little Bustard	195
The Prairie Hen	199
The Willow Ptarmigan	202
The Ptarmigan in Winter Plumage	203
The Quail	205
The Curlew	207
The Golden Plover	210
The Lapwing	213
The Spur-winged Lapwing	214
The Woodcock	217
The Giant Heron	219
The Peacock Heron	220
Group of Herons	221
The Great White Heron	223
The Stork	227
The Marabou Stork	229
The Boat-bill	230
The Spoon-bill	233
Flamingoes	235
The Peacock	237
Peacock Pheasant of Assam	239
The Pelican	241
The Cormorant	245
The Whistling Swan	247
The Black Swan	248
The Black-necked Swan	249
The Wild Duck	251
The Eider Duck	252
The Grey Goose	253
The Spur-winged Goose	254
Stormy Petrels	255

STORIES ABOUT BIRDS.

THE GOLDEN EAGLE.

THE eagle stands at the head of a tribe of great fierce birds—the birds of prey. They may be called the tyrants of their race, for they are constantly seizing and devouring the smaller and weaker birds; and they also attack animals.

Nature has given them great strength of muscle and of talon, and a certain fierce courage that has a kind of grandeur about it. But they lack the intelligence of the smaller birds; they have not the skill of weaving or building those exquisite nests about which we shall speak presently; nor have they the gift of song. Their voices are harsh and screaming; they do not gladden the summer landscape, and their abode is in wild and solitary places.

You see the eagle, as he sits on the crag of some mountain top. He is called the King of the Birds, and well deserves his title. He is monarch of all he surveys. Around him are the mighty peaks of the rocks, the deep dark pine forest, and the chasms, the dells, and the pits, that men behold with wonder and with dread. The eagle's wing has borne him over them with ease. Perhaps he has his nest in yonder ledge of the precipice. There the mother eagle tends her young, and he has come forth to procure them food. He stands erect, the sun glistening on the yellow tints of his plumage until it

THE GOLDEN EAGLE.

shines like gold. He can see far and wide, and deep down below into the valley his glance penetrates.

THE EAGLE.

He is armed for the very purpose of plunder. His beak is hooked and strong, and the edges cut like a knife. His feet have four powerful toes,

THE ROYAL EAGLE, A VARIETY OF THE GOLDEN EAGLE.

armed with sharp talons, long and pointed, and formed for clutching. He has no gizzard, for he requires none. He never feeds on anything but flesh, and the coats of his stomach are firm enough to digest it. He never warbles to his mate, or utters any of those sweet and tender notes that are so pleasant to the ear; his voice is like his nature, harsh and forbidding; it resembles the bark of a dog more than any other sound.

The nest of the eagle is very large indeed, and made of sticks and dead twigs and heath, and it has a hollow place in the middle lined with a little wool and feathers.

The young birds are covered with white down, amid which the feathers are beginning to appear. You can see that the parent eagles have taken care to provide them with abundance of food. The bones of all kinds of small animals lie scattered about the nest, and the half-eaten bodies of grouse and game, the very morsels that are considered to be such delicacies by man.

The golden eagle is the only one of his tribe that lives in Britain, and in the cultivated parts of the island is very rarely seen; he loves wild and solitary places, and the remote parts of the Highlands of Scotland suit him best. The Isle of Orkney is one of his favourite resorts. On one side of the island the sea rushes in with a fury that is scarcely to be equalled in any part of the world; and it has made great rents in the coast, so that there is a line of precipices and caverns that are grand beyond description.

This is just the place for the eagles to dwell. Here they make their nests year after year—or rather fit them up again. The old birds drive off the young ones as soon as they can fly, and keep the nest for themselves.

They are not very pleasant neighbours, as you may suppose, from their habits of plunder.

One day an old minister was walking in his garden, when he heard a loud squeaking noise, that, after being very violent, began to grow fainter. He went to see what was the matter, and arrived at the spot just in time to catch a parting glimpse of his nice fat pig as it was being carried through the air by an eagle.

Another day the eagle, having finished the pig, came again to see what he could find. But this time he made rather a mistake. By way of varying his diet, he swooped on a sheep. But his claws got entangled in the wool, and the sheep was rather too heavy to be carried through the air as the pig had been. The minister had time to get to the spot, and knock the eagle down with a stick.

Of course, the eagles are not at all liked, and the people do all they can

THE IMPERIAL EAGLE.

to destroy them. But it is no easy matter to climb up to the place where the nests are built, and very few persons are bold enough to do it. Sometimes a man is let down over the face of the rock by a rope, till he gets to the ledge, and then he sets the nests on fire.

Sometimes the young eagles are taken out of the nests, and carried away to be tamed.

One of these tame eagles was kept by the squire—or "laird," as he is called—of the district. He used to be chained in the kitchen, where he had rather a happy time of it. The servants made a great fuss with him, especially the cook, who fed him with every dainty.

The eagle was very fond of her; but one day he could not refrain from stealing her shoes. She had left them just within his reach, and he pounced upon them, and thrust his own feet into them. None of the other servants could make him give them up; but when the cook came back, he quietly allowed her to take them off.

THE SEA EAGLE.

It is a grand sight to behold the sea eagle float in the blue sky far above the mountain tops.

He is at home in this higher region—this cloud-land, if we might call it so. Slowly, and with great majesty, he sweeps round in a wide circle, rising and rising until he is no longer to be seen.

His food consists of dead animals, in which respect he is like the vulture. He searches the lonely beach for dead fish, or young sea-birds, and he scours the moors and pastures for what other prey he can find. He does not, like many of his tribe, rise high above his prey. He flies only a few hundred yards above it, and sweeps the hill-sides with outspread wings. Far out at sea the sailors watch him, and he has been seen to clutch at a fish that happened to come to the surface.

But this way of catching fish is now and then fatal to him. The fish, if it is a large one, contrives to pull the eagle down under water, and then he is drowned. Should he escape such a fate, he keeps fast hold of the fish, and, half opening his wings, brings it to the shore.

Then he takes care to get his claws at liberty, and to dry his feathers, so that he can fly at a moment's notice; after which he quietly begins his repast.

The eagle does not despise the bank of a river or a lake, for here he can now and then feed delicately on salmon and trout.

THE SEA EAGLE.

He often sees the otter catching a fish, and he waits until the creature is satisfied, and takes what is left. For his courage is not quite equal to his size and his strength, and he rarely attacks an animal larger than a hare. Indeed, he has a touch of the vulture about him, and will eat dead creatures with more relish than living prey.

In order to escape from his greatest enemy—man—the eagle has retreated to the wildest and most desolate part of the coast, and makes his nest where scarce any living creature can reach it. His nest is of immense size, and is made of sticks and heath and twigs and dead sea-weeds. The mother bird lays two eggs, of a pure white, with some pale red dots at the larger end. The young birds are clothed with a greyish-coloured down, and are plentifully supplied with food. But as soon as they are old enough, the parent eagles drive them away.

The shepherds and farmers in the neighbourhood have a great dislike to the eagle, and try to kill the young ones. They contrive to creep along some mountain track till they get to the nest, and then, as in the case of the other eagle, set it on fire.

During this process the parent birds wheel round and round, and utter screams of distress. They might easily attack their persecutors and drive them away, but they rarely attempt to do so. Yet now and then a person crossing the lonely moors has been scratched and buffeted by an eagle.

There is another fierce bird of prey, called the osprey, or the fish-hawk, that is a distant relation of the eagle. He, too, feeds on fish, and hovers over the water in a hawk-like fashion. Then, when a fish comes near enough to the surface, he pounces on it, and is seen rising with it in his talons. He thinks he is secure of his prey, but now and then he meets with a disappointment. The sea eagle has been watching close by on some crag on the lonely beach. Now he bends his head, makes a great swoop on the hawk, and frightens him so that he drops his prize. Then the eagle, by a very adroit swoop, catches it before it reaches the water, and carries it off.

There is a story about the sea eagle that some people can hardly believe, though others declare it is true.

He is said to wet his plumage in the sea, and then roll about on the sand until a great deal of it adheres to him.

It is in Norway where this happens, and at a place not far from some mountain pasture, in which cattle are feeding. He does not attack the poor ox in an open manner, because, as I told you, he was rather a coward. He hovers over the ox, and by-and-by begins to shake the sand from his wings,

THE TUFTED EAGLE.

so that a great deal of it falls into the eyes of his intended victim. The ox is blinded, and also begins to get frightened by the noise of wings over his head. He runs about in a kind of panic, and as the noise goes on, and the sand keeps falling, he loses all sight and sense. There is generally a steep place like a precipice at the edge of the field, and the ox is almost sure, sooner or later, to run over it and be killed. Then the eagle can easily slip down and devour him.

THE TUFTED EAGLE.

THERE is a small eagle with a tuft on his head that lives in Africa, and has such a very dreamy appearance as he sits on the branch of a tree, that you think he is asleep, or else is a very stupid bird.

He lives in those fertile parts of the country through which the river Nile flows and gladdens the scene. There are fields, and villages, and beautiful groves of the mimosa-tree. He often sits perched on a branch for hours together; but he is not asleep, as you might suppose. Only watch a few minutes and see what will happen. Yonder is a little squirrel playing blithely among the branches. It is as happy as can be, and runs merrily about, as if it feared no evil.

But by-and-by, in its gambols, it ventures near to where the eagle is sitting. The eagle has seen it all along, only he pretended not to do so. He did not want to frighten the squirrel away, but to get it into his clutches. Now the right moment has come. He rises, raises his wings, and gives a terrible pounce. You would not have thought he had been so strong or so fierce. But it is all over with the poor squirrel.

Sometimes the eagle plays the same game with a mouse or a rat, or any little bird that, in its happy freedom and joy of heart, ventures heedlessly near the fatal spot.

His eyes have a fiery expression, and are a bright yellow colour; and his plumage is brown. His nest is in some tree, and is lined with feathers. Though he is small, he is as savage as any of his tribe; but if he is kept in confinement, he becomes rather cowardly, and loses his ancient spirit.

PEREGRINE FALCON.

In olden times, in "merrie England," as it used to be called, many ancient sports were carried on that have long since passed away.

PEREGRINE FALCONS.

One of these was hawking by means of a race of birds called falcons. The falcon used to be blindfolded, and fastened by a chain to the wrist of its owner. He was then carried out into the fields, and when a wild fowl, or heron, or any suitable prey, was seen, the bird was unhooded, and let fly. The amusement, which was rather a cruel one, consisted in seeing the falcon strike down its prey. The art of falconry, or hawking, was such a fashionable amusement, that people of rank hardly ever stirred out without their hawks perched on their wrists; and a man, called a falconer, was employed to take care of them and feed them.

The peregrine falcon is of a family that once stood very high in the public esteem. His ancestors were used in hawking, and were fed and caressed by kings and nobles. No one was allowed to injure them, or to meddle with their nests. But times have changed since then. The descendants of those highly favoured hawks are now in little esteem. It is true they possess the same qualities. They are as bold, and as brave, and their sight is as keen as ever; their plumage is as handsome, and they could be taught just as readily. But fashions are altered, and no one wants them. They are now despised and persecuted, and may be shot with impunity, or their nests rifled without the least danger.

In the old days of hawking, the bird that was chosen for the purpose must possess certain qualities, all of which were united in the peregrine falcon. He was full of spirit and daring, and would attack any bird without hesitation; and he had great strength of muscle, and was able to contend with the larger kind of game; and, joined with these qualities, he was very obedient to his master, came at the word of command, and could be petted and caressed by those who had the care of him. In fact, the falcon was called "noble," because he was the noblest of all the birds of prey.

His shape is very compact, with a full, well-rounded breast, short neck, and large head; his bill is short and thick; and his eyes are large and keen, and of a deep hazel colour; his claws are, as you see, very strong and well curved, and able to grasp and seize almost anything; his plumage is dense and strong, more so than that of the rest of his family. Its general colour is deep bluish grey, and the wings are barred with black; the throat and neck are white. The mother bird is the larger of the two, and her plumage is redder on the lower part and less blue on the upper. In the old days she was always called the "falcon," and her mate the "tercel," and he was flown at the smaller game, such as partridges and magpies. The peregrine falcon is often seen in the wild moors of Scotland, for here he finds plenty of grouse and partridges and rabbits; and occasionally he falls upon a young gull—that is, during his visits to the coast.

He flies swiftly, and does not often balance himself in the air as the eagle does, for his wings are shorter. When he sees his prey he pounces upon it in a slanting direction. He is a silent bird, except when he has a family to care for, and that makes him anxious and excited, and he utters now and then a clear shrill cry.

He is extremely daring when he is hungry, as you may think from a little anecdote I can tell you.

One day a sportsman on the moors saw a falcon hovering over a grouse, and following it about. By-and-by he dropped down upon it, and when the sportsman came up with his dogs, the falcon was devouring his prey. His partner was close by, feeding on another grouse. Of course, the birds were obliged to rise, but they did so very unwillingly, and barely kept out of the way of the dogs. Meanwhile, the dogs started several grouse, and the falcons must have thought it was done for their special benefit. They pounced on them every one, and struck them down under the eyes of the sportsman, and as if in defiance of him.

The nest of the falcon is on the face of a cliff, and, as a rule, beyond the reach of man. It is very large, and is made of sticks mixed with the stems of grasses. There are three or four eggs in the nest, of a dull red colour, with dark spots. The young falcons are supplied with abundance of food. The parent birds can bring them pheasants and pigeons and plovers, and all the delicacies of the season.

In the islands of Shetland the peregrine falcon chooses the most rocky and desolate spots. He and his partner make their nest on the face of very high cliffs. No other falcon seems inclined to live in the same cliff, for the birds are not social. But the different kinds of gulls and sea-birds crowd the cliff, and build their nests in every ledge.

The falcon does not regret this arrangement in the least. He considers a young gull to be a dainty morsel, and he is sure of getting a great many. He waits till the old birds are out of the way, and then drops suddenly down on the nest, and carries off the little gull he has been longing for.

THE GOSHAWK.

THE hawks may be said to be cousins to the falcons. They are moderately-sized birds, and occupy a position between the falcons and the buzzards.

THE GOSHAWK.

Their bodies are rather slender, their wings short and round, while the tail is long. They fly low when they are searching for prey, and have a gliding and stealthy manner. They nestle in trees or on rocks, and their nest is a little like that of the crow.

There are but two species in England. One is the goshawk, the bird in the picture, that is so rare it is hardly ever met with; and the other is the bold, hardy, and rather insolent sparrow-hawk, which is as common as his relation the goshawk is rare.

The goshawk may be known from his cousins the falcons by the curve of the upper part of his bill, which is peculiar to the hawk family. He is not

THE SPARROW-HAWK.

so strong as they are, though he is quite as large; but his way of flying at his prey is different. He does not make a swoop, but glides along in a line after it—a mode of proceeding which, in the language of falconry, is called "raking."

The goshawk used to be in great favour in those hawking days, and was let fly at pheasants and hares and partridges. He used to make a great dash, and even go through a wood or thicket after the prey. But if he did not soon catch it, he seemed to grow tired, and give up the pursuit. He would then perch on some bough of a tree, and wait patiently till a new victim appeared, or until the old one came again within his range. If the game lay hidden close by, the goshawk, perched on his bough, would wait for it to move. He would remain motionless for hours and hours, until the intended victim was forced by hunger to come out of its lair; then the hawk would pounce down upon it.

The nest of the goshawk is placed in some high tree on the outskirts of a wood, and the birds will occupy it year after year. There are three or four eggs, of a pale bluish white.

THE SECRETARY BIRD.

THIS very warlike-looking bird might, at first sight, be thought to belong to the tribe of long-legged storks or cranes. But if you examine his curved beak, you will see that in reality he is a bird of prey. Indeed, some people call him the "secretary eagle."

The reason why the name "secretary" has been given him is because of the crest of feathers on the back of his head, that have a fancied resemblance to a pen stuck behind the ear of a person employed in writing. But he might be said to have a link with another family of birds, namely, the running birds. He cannot grasp like the eagle, and he does not live, like his noble relative, on high mountains, or soar towards the clouds. On the contrary, he keeps on the ground, and runs here and there on his long legs. So that you perceive it is rather a difficult matter to find out where to place him among our feathered friends.

He is one of the most useful birds I can name, and in certain parts of the world is cherished with the utmost care. I need not tell you that he does not live in England; and if he did, there would be very little for him to eat. Happily we have not many snakes, except a few small harmless ones; for though you will remind me of the viper, it is very rare, and hardly ever seen except in woods and solitary places. At any rate, we have none of those great serpents that abound in the places where the secretary bird lives. He does not object to lizards, and even beetles, by way of variety; and as he runs about on the hot, dusty plains of Africa, he finds plenty. But this is child's play; he likes best of all to do battle with a serpent.

Many venomous snakes are found in these hot countries, and the natives dread them beyond measure. It is true the snake will rarely attack a man, and, as a rule, glides away from him; but sometimes he may chance to come

THE SECRETARY BIRD.

too near it, as it lies coiled up, and if its terrible fangs do but touch him, he is sure to die speedily. And there are many stories told of snakes that by mistake have got into a house, and even nestled under a pillow.

The secretary bird is always on the look-out for this natural enemy of man. In the picture he is engaged in a fierce battle with a serpent. The serpent is, as you see, in a rage. At first all its attempts were directed to getting back to its hole, but its enemy was more than a match for it. Whichever way it turned the bird hopped just in its path, and stood with flashing eyes and outspread wings. Then the serpent was fairly roused. It raised itself up, swelled out its dreadful neck, and darted out its fangs. For a moment the bird gave way a little, and seemed as if considering what to do.

But his courage soon revived. He was resolved not to be cheated of his prey, so he covered himself with one wing as with a shield, and struck violently at the serpent with the other. The serpent was knocked down by the blow, and every time it attempted to rise, the bird struck at it again. At last the snake could rise no more, and then the bird killed it by striking its head with his beak.

These kind of battles are often taking place, and the bird is much admired for his courage. He is considered a most valuable member of society, and his family have been invited over to the plantations in the West Indies. Here they are highly esteemed, and no one ever thinks of harming them. The plantations abound in snakes, and their number is thinned by the introduction of these their inveterate enemies.

When the snake is small enough, the bird snaps it up, and carries it off to the top of a tree. Then he lets it drop, and follows it, as it descends, with much adroitness, so as to be ready to strike it when it lies stunned on the ground. He does not always strike with his wing, but with the sole of his foot. He always kills his prey before he devours it.

Serpents are not his only food, for he preys upon lizards and tortoises and insects. The hot unwholesome marsh is full of insects, and the secretary bird thins their number; so that every way he is useful.

He and his partner make a large nest, in which two eggs are laid. He does not choose his partner without fighting a great many battles. Yet he has not at all a fierce temper, but rather otherwise; and after the choice has been made there are no more quarrels.

THE VULTURE.

ALMOST every one dislikes the vulture. His very name is thought to express cowardice and gluttony, and every unpleasant quality. He is not to be com-

THE VULTURE.

pared to the eagle and the other birds of prey, and his place hardly seems to be among them.

His head and neck are bare of feathers, and his plumage is coarse and ill kept; his eyes are prominent, and his claws shorter. Indeed, he cannot carry his prey through the air as the eagle does, but is obliged to stay and devour it on the spot.

His prey consists of dead creatures, and he has not the courage to attack living ones. Unless, indeed—and this fact shows his cowardly nature—unless the poor creature is either wounded or dying, and can offer no resistance.

His family are spread all over the world, but are more abundant in hot countries than in cold ones. Happily we have no vultures in England, and none are wanted.

They are wanted in many places, for people in tropical towns and cities have not very neat habits. They allow heaps of rubbish to lie about in the streets, and dead creatures are thrown there in a way that would not be tolerated in England. But the disagreeable vulture does not object to act as a scavenger; it is in accordance with his nature; and he and his companions busy themselves in clearing away the rubbish, that, if it were left, would soon cause a fever. In this respect the bird is useful.

In South America there are vast herds of wild cattle, that are the riches of the inhabitants. Many of them, nay, even thousands, are driven every day to the town to be killed and then salted, and packed up and sent away as an article of food. Of course, this kind of business always being carried on, would be very unpleasant to the inhabitants, for the streets become littered all over with refuse parts of the meat. But the town is full of vultures, that are as tame as domestic poultry. No one meddles with them, and they employ themselves all day long in clearing away the decaying matter into their own greedy maws. So that the town is kept quite neat and clean, compared to what it would be. Here again the bird may be said to be useful.

And in the open country, and on the mountains and sea-shore, and, indeed, everywhere, are the vultures plying their trade.

If a camel falls on the desert, or a mule in the passes of the mountains, its body is not left to decay and to poison the air. As if by magic, a vulture is sure to appear, and then another, and another. In an hour, nothing will be left but a heap of bones, picked as clean and as bare as possible.

The eagle sometimes condescends to preside at these feasts, and then the vultures keep at a distance until he is satisfied. After every morsel has been eaten, the vultures will be so full that they cannot rise, and can only hop along the ground.

There is one of the family, called the white vulture, that has a great fancy for the eggs of the crocodile. You would think he had very little chance of getting any of them, guarded as they are by the fierce mother crocodile.

But all the while she was burying them in the sand, according to her usual habit, the vultures were watching her. By-and-by she finished her task, scraped the sand over the place, and went away for a little diversion, perhaps to wallow in the mud by the river-side.

THE TAWNY GOOSE VULTURE.

The moment she was out of sight, the vultures, for there were several of them, began to bestir themselves. They uttered loud cries, and pouncing down on the nest, began to scrape away the sand, and devour the eggs. The same vulture contrives to get a taste of the eggs of the ostrich. The natives declare that when the parent birds are away from the nest, a stone is seen to fall into it as if from the sky. But, in reality, the vulture has dropped it from where he is hovering high up over the nest, on purpose to break the eggs. Then down he comes, and feasts upon them to his heart's content.

The tawny goose vultures are met with in almost every part of Africa, and are very tame. They have the useful habits of their tribe, and clear away rubbish in a very short space of time.

THE CONDOR.

THE grandest of all the birds of prey is the condor. He can soar higher than the eagle, and though he belongs to the tribe of vultures, that are an inferior race to the king of birds, yet he excels him in size, in strength, and in fierceness. Indeed, none in the whole family of birds can compare or compete with him.

The size of the condor is immense. When the wings are spread out they measure as much as twelve or fourteen feet across.

He loves the mountain ranges of the Andes; and here the solitary traveller often sees him, soaring aloft until he becomes a mere speck, and entering a region where man could not breathe the rarefied air.

At night the condor rests on the ledges of the rock; and when the sun gilds the mountain tops, and while it is yet dark in the valleys, he rouses himself, flaps his wings, and peers over the ledge into the abyss below; then he dives over, and seems as if sinking by the great weight of his body; but soon he rises, and begins to move upwards in sweeping circles, until he ascends often as much as four miles above the level of the sea.

By-and-by he descends to the shore, and his loud screech is heard with the dashing of the surf. When hovering in the air he will spy out his prey in the valley below. Sometimes it is a lamb, or a sheep, or a mule that has fallen dead on the mountains.

True to his nature as a vulture, he will not reject dead prey, but he is equally glad of it even when alive.

His talons cannot clutch the prey as do those of the eagle, and he does not attempt to bear it aloft in the same way as the royal bird; he is obliged to eat it on the spot.

THE CONDOR.

He fixes it, as it were, to the ground with his claws, and then rends and devours it with his beak.

c

Like the rest of his race, he is a great glutton, and will feed until he is unable to rise again into the air. He may then be approached, but rather at your peril, since he fights desperately, and is more difficult to kill than almost any creature in the world.

We can tell you an anecdote about the condor's power of life.

A miner in Chili, a very strong man, once saw a condor enjoying his feast on the mountains. He had eaten so much that he could not fly, and the man attacked and tried to kill him. The battle lasted a long time, and the man was nearly exhausted. But in the end he thought he was the victor, and left the condor dead, as he imagined, on the field. Some of the feathers he carried off in triumph to show to his companions, and told them he had never fought so fierce a battle. The other miners went to look at the condor, when, to their surprise, he was standing erect, flapping his wings, in order to fly away.

A bird with such powers of life continues to exist years and years. Indeed, the condor is said to live for a century.

The Indian tries to catch the condor by stratagem. He employs him to fight in a ring, at those cruel bull-fights which are the favourite amusements in that part of the world.

He does not attempt to attack the condor openly, for he knows how strong he is, and he wishes, besides, to take him alive.

He procures the skin of a cow, and hides himself beneath it. Some pieces of flesh are left hanging to the skin, and are sure to attract the condor. He comes pouncing on the prey, and while he is feeding with his usual greediness the Indian contrives to fasten his legs to the skin. When this is done, he comes out of his concealment, and the bird sees him for the first time. He flaps his wings, and would fly but that his feet are entangled; and, more than this, a number of other Indians come running up, and throw their mantles over him.

There is another way of taking the condor, but without saving his life.

In a certain place in the mountains there is a large, deep chasm, that might be almost called a condor trap.

A dead mule is placed on the brink of the chasm, and very soon the condors scent it out. Down they come, and soon pull the mule over into the abyss. They follow with haste, and feast until they can scarcely stir. Then the Indians follow with sticks, and kill as many as they can.

The condor does not trouble to make a nest, but chooses a hollow in the cliff, where the mother bird lays her eggs. Both parents devote themselves to the young ones, and feed them with the utmost care and attention. The young birds grow very slowly, but at the end of six weeks they begin to flutter round their parents. After a few months they fly off to seek their own fortune.

The condor is a handsome bird. His shining black feathers are two feet in length. His bill is very thick, and the point hooks downwards; it is white at the tip, and the other part is a jet black. A short down covers the head; and the feathers on the breast, neck, and wings are of a light brown.

THE KITE.

HAVE you ever, on a summer's day, seen a bird gliding about in a circle, with outspread wings and extended tail?

His way of flying was like that of the eagle, and yet he was a much smaller bird. Sometimes he balanced himself in the air, and ceased to move, but hung suspended, as it were, on nothing. Then, while you were still looking, he glided downwards to the ground.

While poised in the air, his keen eye had been fixed on some object below; for he seeks his food on the ground, and is very quick at spying it out. Lizards, frogs, mice, and even young birds, fall into his clutches: nothing comes amiss to the kite. He is a bird of prey as much as the eagle, only that he has not the strength or the bravery of the king of the birds; and he descends to acts of theft and violence, like the whole of the tribe, great or small. He is often hovering over the farmyard, and if the hen does not take care of her chickens, he is pretty certain to carry some of them off.

But he is a sad coward; and if the hen sees him, and comes rushing out, as she always does under the circumstances, looking angry and excited, and with her feathers ruffled, he never attempts to withstand her. He would do anything rather than fight, and she drives him away as easily as possible.

The brave little sparrow-hawk, a relative, as I told you, of the goshawk, can do anything he likes with him. He is the smallest of the two, and, we must confess, has a very quarrelsome temper; but the kite flies away as fast as he can to get out of his way. He seems as if he would hide himself in the very clouds. But the sparrow-hawk is bent on finishing the squabble, and he goes after him, and beats and buffets him, till the kite is often brought to the ground, more frightened, perhaps, than hurt.

The kite often gets into trouble by his love for young ducks and chickens. He is so intent on his act of plunder, and so greedily desirous of his prey, that he forgets his usual caution. A hen-coop once stood in a farmyard, and the young ducklings, which a hen had hatched, were waddling about and enjoying themselves. The kite saw what was going on, and knew that the hen could not interpose; so he made a great pounce upon a poor little duckling. The duckling screamed loudly, and ran to the pond for safety. The kite followed as close as could be, and even ventured into the pond after it. In the meantime the servant-girl had heard the screams, and went to see what was the matter. The kite could not fly all at once, and the girl had time to knock him over with a broom.

In the engraving, the kite is sitting on the bough of a tree, as if he had had his repast, and were quite content.

You can see from his hooked beak and sharp talons that he is related to the eagle and the hawk. His beak is short, and his plumage soft and glossy. Truly he is a handsome bird; and if he were to leave his bough, you would see that his manner of flying is very graceful. His wings are long, and his tail is forked; he glides and wheels about in all kinds of attitudes, and it is a pretty sight to watch him.

In some parts of Scotland he is very rare, and it is quite a matter of surprise to see a kite. Once a kite used to make his appearance every day about the same hour, and wheel, and curve, and go through all his evolutions, to the great delight of the people who saw him. They would stand and watch him in wonder and admiration. A naturalist tried in vain to catch the bird, and followed him over moor and mountain. The kite always kept out of his way; but as he glided along he would swoop now on a partridge, and now a grouse. Indeed, he made great havoc among the game. His love of game cost him his life; for one day he was so intent on devouring a partridge that he did not see his enemy stealing along towards him; and then the cruel gun was raised, and down dropped the poor kite.

The nest of the kite is made of sticks, and lined with feathers and down.

THE COMMON KITE.

It is placed in the fork of a tree in some deep wood. The eggs are of a dirty white, with a few red-brown spots.

When his nest is attacked, he becomes very bold, and will scream and go into a great excitement; and he will attack the intruder, and try to drive him away.

Two young kites were once taken from the nest and tamed. They became very friendly and affectionate. Every morning they were allowed to fly at liberty, and they used to soar in the air, and make a great many of their graceful curves. But they never went far from home, and always came back to their owner.

There is a kite in America, called the Mississippi kite, that is a bird of passage, and makes his appearance with the swallows.

He comes sailing up the grand and mighty river after which he is named, sweeping along, now with the wind, and now against it. All the time he is looking out for prey. The air is full of giddy whirling insects, that have come out to enjoy the sun, and he swallows some of them in passing. His quick eye notes every object on the bank. Yonder is something that shines with a green colour changing to brown. It is the throat of a lizard that has climbed up the trunk of a tree to look out for food. It feels by a kind of instinct that the kite is near, and is struck motionless with fear. It does not attempt to move, but seems as if it were turned to stone. Another moment the kite has pounced upon it, and it is seen no more.

Sometimes the kites in that part of the world assemble in a group, and sweep round and round a tree. They are catching the locusts that are beginning to be a scourge, and for this very useful act one should think the inhabitants ought to be grateful.

The kites build their nest in the thick boughs of the bay-tree, and here they rear their young. They are devoted parents, and feed and tend the little kites until they are able to get their own living.

A mother kite once took her young one in her claws, and carried it out of the reach of a gun that was being fired from below.

THE BUZZARD.

THE buzzard is a relation of the falcon, as you can see by the curve in his beak. And you can tell pretty much what kind of a life he leads by the attitude in which he stands. He has the same habits as the kite, and has been

THE BUZZARD AND ITS PREY.

flying over and over the ground in a slow and steady manner, looking for prey. He eats worms and mice, and even little birds when he can get them. On this occasion he spied a rat running along the ground, and at once struck it down with his talons. Like the hawk, he kills with a single blow.

THE COMMON, OR MOUSE BUZZARD.

When he has finished his meal, he will retire to the branch of a tree, or to the ledge of a rock, and repose until he has digested it. Then he will issue forth again, and sweep over the country, or sail about with his wings expanded to their fullest extent. Then he looks very much like the eagle.

His nest is built in a tree, or else on the shelf of a rock. It is made of sticks and twigs and heath, and has a rude lining of wool. While the mother bird is sitting, her partner is very attentive to her wants, and is constantly bringing her food.

In some parts of Scotland the buzzard is very rare, though we often see him in England.

But now and then the Scotch farmer sees a bird hunting about in the fields, with a wavering, uncertain flight, and making all kinds of turns and twists. Soon he observes it drop down suddenly on some unwary prey beneath, and he knows that it is the buzzard. He ought to feel a friendship for the buzzard, instead of shooting him. The mice nibble the corn, and the more they are seized the better. And when the corn is ripe, a number of pretty ring-doves will come in flocks to eat it. The buzzard would have caught a great many of them, if the farmer had not shot him.

THE SNAKE BUZZARD.

LOOK at the picture in the next page, and you will see that the snake that hangs from the bough of the tree has just been killed by a very useful bird—the snake buzzard.

In countries where these venomous reptiles abound, it is of the utmost importance that their numbers should be thinned. It is true the snake, as a rule, has a dread of man, and will bustle away among the dead leaves on the ground when he hears footsteps approaching. But many fatal accidents are always occurring, for the snake may be trodden on without being seen as he is coiled up asleep, or he may creep into the house. A lady once felt something move under her pillow in the night. She took no notice, but the next morning, when the pillow was moved, there was a venomous snake coiled round, and it had lain there all night.

Some of the birds of prey seem to have an especial desire to kill snakes. Besides the brave secretary bird, the buzzard of which we are speaking does his part to keep them down. His family live in many parts of the world—in Europe, Asia, and Africa. He has great eyes, that shine like those of the cat or the tiger in the dark. But he is not very fierce, though he quarrels very often with his companion buzzards. He has a way of hopping about a little

THE SNAKE BUZZARD.

like a raven, and he likes to sit on a solitary branch, as you see he is doing in the picture, and to take a survey of all around him.

It is a good thing that his feathers are so thick and close. They are like a suit of armour, and defend him from the bite of the snake. He does not mind what kind of snake it is, or how poisonous. He darts down and attacks it at once.

His way of attack is by seizing the snake with his claws just on the nape of the neck. The snake is struck down, as it were, and cannot use its fangs. It often twists the rest of its body about, and wraps it quite round its enemy. But it cannot move its head, so there is no harm done; and the buzzard gives it a great bite, and ends by killing it. He eats the snake bit by bit until none of it is left.

You must not suppose that he lives entirely on snakes, though he kills as many as he can, and will eat three great ones in the course of a morning. He does not object to rats and mice and other small animals, and he rather likes to fish in shallow pools, and has no objection to a crab, whenever he can find one. He is rather handsome in a suit of brown, his tail tipped with white, and the under part of his body white, with brown spots. He and his partner make a nest of young twigs of the trees, and line it with leaves. It is a little the shape of a saucer.

THE SNOWY OWL.

THE owl is one of the birds that is very rarely seen. The reason is because of his secluded habits and his dislike to facing the light. It must be some very unusual circumstance that can bring him out in the day-time.

A gardener was once working in a garden when he heard a very strange noise from the top of a tree. As he was very expert, he climbed up to see where the noise came from, and what it was that made it. When he got half way up the tree, two fierce white creatures dashed out and attacked him with beak and claws, making at the same time a terrible screaming.

They were, as the intruder soon found to his cost, a pair of owls taking care of their young in a nest at the top of the tree. And an owl in a passion is no pleasant object to meet with.

THE SNOWY OWL.

The man hurried down as fast as he could, but he had some difficulty in keeping off the owls. In spite of the daylight, they darted at him again and again, wheeled round his head, and even pursued him, much scratched and frightened, to the very door of his retreat.

But, as a rule, the owl lies very safe and snug in his roost, and does not stir till twilight. He is a very curious bird, and we must spend a few minutes in making rather a close acquaintance with him.

He is a bird of prey, for he hunts mice and rats, and even small birds if they chance to be about in the twilight. His feet are formed on purpose to grasp the prey. The toes are feathered; the first toe is the shortest, and the fourth toe is longer and can be turned backwards. The claws are long and curved, and very sharp.

Do you notice the thick plumage of the owl? It is as fine and soft as possible; and when he drops from the branch of a tree to the ground, which he sometimes does, when he chances to spy a poor little mouse moving beneath, he makes no sound: the mouse cannot tell he is there until it feels the sharp talons.

On the side of the head are some loose, slender feathers that make a circle, or, as it is called, a disc. And sometimes the owl has a tuft of upright feathers on each side of the head, like horns, and then he belongs to a family called horned owls.

The eyes of the owl are very large, and the circle of feathers round them reflect the light upon them as a reflector does upon a lamp. But the worst of it is that these large eyes take in more light than the owl can bear. In the day-time he is blinded by the excess of light. This makes him appear as if he were stupid, and he blunders about as though he had lost his senses.

The little birds hate the owl, for he pounces upon them whenever he can, and many of their companions have felt his cruel claws. It is fine sport to them if, by any mistake, the owl chances to be abroad in the day-time. They soon find it out, for one tells the other, and there is quite an uproar in the garden.

It is never generous to take advantage of a defenceless enemy; but the little birds do not think of this. They have many wrongs to revenge, and they fly at his face, and even peck him, taking care, however, to keep away from his claws, and they scold, and drive him about to their hearts' content.

As a rule, he does not try to defend himself, but flutters dizzily about, and goggles with his great eyes. But if he stops and turns round upon them, the rabble rout at his heels take to flight in a moment.

HABITS OF THE SNOWY OWL.

But the eyes of the owl, though they do not help him much in the day-time, are of the utmost service in the twilight. He can see the smallest speck on the ground, or the tiny mouse in the corner of the barn. And the farmer rather likes him on this account. One barn owl is as good, and will do as much work, as a dozen cats.

But as there is no rule without an exception, so there are owls that can see by daylight. The snowy owl is one of these. You see he is not hiding in a corner, but sits very composedly on the bough of a tree, and does not wink or blink although it is day-time.

He lives in the northern parts of Europe, and goes southward when winter approaches.

His flight is noiseless, like that of the other owls, but he can continue on the wing for a long time. Sometimes he hunts in the air. He spies a pigeon or a wild duck, and he sets himself to follow it. With his swift and steady flight he soon gains upon it. Then he strikes it with his talons, a little in the same manner as the hawks do.

He loves the margin of rivers or streams, and if there is a rapid, or a waterfall, he is all the better pleased. There he stations himself, for plenty of fishes are sure to be drawn over, and then he pounces upon them. He also goes to the trap in which some small animal, such as the rat, is caught, and devours it. His diet consists also of larger prey, such as hares and squirrels, and his meals are excessive. You would wonder how his stomach could hold the amount of food put into it. But, happily, it has the power of stretching out like india-rubber, which exactly suits him.

If the snowy owl could not hunt in the day-time, I do not know what would become of him. For in those northern regions, at certain seasons of the year, there is no night at all. When he is taken alive and kept in captivity, he is very gentle, but easily alarmed. Then he raises his head, opens his mouth, and utters a sharp low cry.

The owls build their nests on steep rocks, or among the branches of the pine. The mother bird lays two eggs of a pure white. Sometimes it is said that the nest is placed on the ground.

A ship was once sailing to America, when the people on board saw a snowy owl far out at sea and skimming the waves. Though he must have been a long time on the wing, he did not seem at all tired, but rose and fell with the crested billows as though he enjoyed the sport.

When the vessel returned to England, a naturalist who was on board had a snowy owl as a pet. He had bought him of an Indian, and kept him during the voyage.

At first the owl was very timid, and tried to get away if any one came near him. But by degrees he became more courageous, and devoured the pieces of meat given to him with an excellent appetite.

His disposition was very gentle, and he never attempted to strike with his claws, or to bite anybody.

The captain put up a roost for the owl, and used to amuse himself by shaking hands, as he called it. This was done by putting one of his fingers among the talons of the bird and shaking the foot, often very roughly. But the owl seemed highly delighted, and used to support himself on the other leg. He was a great favourite with the sailors, and they used to give him all the scraps of meat they could. Indeed, he was so used to be fed that he would take the bits from them when dozing on his perch, and without being at the trouble of opening his eyes.

One day, as he was having a nap, a sailor held his hands full of salt water before him. The owl thought the water was meant for a bath, and dropped his head into it, a mistake that was very unpleasant.

The next time the trick was played the owl knew better, and gave the sailor a great bite.

One day at dinner, when the sea was very rough, the owl was thrown from his perch, and alighted on the bald head of the carpenter, who was sitting at the table. The carpenter pushed him off, and he flew to the mate, and settled himself firmly on his thick bushy hair. Nor did he seem in any hurry to depart, for though the mate kept striking at him, every blow was repaid with interest by a hard grip of the talons, and a box on each ear from the two powerful wings.

At length, however, when the sea became calmer and the vessel ceased to rock, the owl went away to his perch.

The plumage of this owl, as you may suppose from his name, is white, and in the winter can hardly be distinguished from the snow. The only part of our own islands that he inhabits is Shetland, and once or twice he has been seen in Orkney. He likes solitary districts, and in these islands does not, as in America, come abroad in the day-time.

On the approach of twilight, he prowls over the fields in search of mice and small birds. When he first leaves his retreat, the crows and other birds attack him. But he seems more amused by their clamour than alarmed at it, and dashing through the air soon leaves them behind.

He is considered a bird of ill omen, and few people like to meddle with him.

THE SHORT-EARED OWL.

THE SHORT-EARED OWL.

Now and then the sportsman, as he makes his way over the fields in the south of England, rouses a bird that was sitting very contentedly by the green margin of a stream or brook.

He was just in the position in which you see him in the picture, and appeared in a half dozing condition. But on being disturbed he flew away, darting hither and thither in different directions. When he had proceeded in this way a few hundred yards, he stopped and settled down again.

It is not clearly known whether he hunts by daylight, but, at any rate, he has rather peculiar habits, considering he is an owl. He is often seen among turnips, or by the side of a hedge, and even among the long grass.

He is called the short-eared owl, and he has, as you see, two little tufts on his head, though they do not show much. His eyes are surrounded by brownish-black feathers, that give him a very sleepy appearance. He is never seen to perch on a tree, but generally hides in the grass, and likes open barren situations. His nest even is placed on the ground, or among reeds and rushes. One nest was found upon a moor or common. It was made by scooping a hole, and then the eggs were placed in it without any further preparation. In America a nest has been seen on a mountain ridge, and another under a bush. It was built in a slovenly way, of dry grass raked together, and was very large.

The mother owl was sitting on her eggs, and would not have been noticed if she had not made a curious clucking noise with her bill. She was so intent on her business of hatching, that she allowed herself to be touched before she hopped away. And then she went a very little distance, and came back as soon as the intruder was gone.

THE STONE OWL.

THE stone owl looks so very wise that he has been called "Minerva's owl." For Minerva, as I dare say some of you know, was worshipped by the ancient Greeks as the goddess of wisdom. And the Greeks knew the stone owl very well, and used to paint him in the picture of their goddess.

His body is dark grey, spotted with white; and his knowing little face is a greyish white. In the picture he looks as if he had been studying some hard problem, and were coming out of his lair to give us the benefit of his learning.

He has one habit that has made him rather dreaded by ignorant people.

He likes to hide in an old wall, or behind a tomb in a churchyard. And from out the darkness he will give a very unpleasant screech as he issues forth to seek his food. Like the rest of his tribe, he flies without making a sound; but the peasant who is hurrying home hears the screech, and pretends that the bird is telling him that some misfortune will happen; so he dislikes the stone owl very much.

But the family of stone owls are met with in many parts of the world.

THE STONE OWL.

They live very happily together, and often go in a body in search of food. Though they fly by night, they can see well in the day-time; and they are very useful in clearing the houses of rats and mice.

We do not mean English houses, but the dwellings of workmen and artisans in Italy. They eat up the mice and insects that infest the place, and in return their master feeds them, and is very kind to them. The wise little owls often sit in a row over his head while he is at work.

A near relative of Minerva's owl is met with abroad, and has a very curious history. He is as much like the bird we have been speaking of as can be, but he has a different name. He is called the burrowing owl, or the mouse owl; but this term is not strictly correct. He does not burrow for himself, but lives in the burrows that have been already made by the marmot. He is sitting now at the entrance of the marmot's hole as if it belonged to him.

He is almost always found there. The marmots and the owls seem as if they lived together, but they do not. It is only when an enemy comes near that they both squat down under shelter together. But they frisk about in company, so that at a distance they are often taken one for the other.

The marmot is often called the "prairie dog," and the burrows it makes, that are very large, and extend a great way, go by the name of "prairie village."

In some parts of North America the "prairie village" reaches under ground for miles together.

At any rate, the owl has quite discarded the habits of his family. He lives in open daylight, and hunts about in the full blaze of the sun. He has no kind of intimacy with the marmots, as some people used to think from seeing him so much in their company. He usually selects a burrow that has been deserted, and his nest, that is made of grass, is placed at the entrance of the hole. Four white eggs are laid in the nest; and the food of the young ones is grasshoppers, or crickets, or even field mice.

The owl often perches on a bush over his hole for a long time together. When any one comes near, he makes a low chattering sound, and skims along the plain. If he is attacked on the wing, he makes for the nearest burrow, and takes shelter there. Then it is no easy matter to dislodge him.

In the winter the owls retreat to their burrow, and are said by the Indians to fall into a torpid state. At any rate, their burrow is in a very different condition from that of the marmot. It is ruinous and neglected, while that of the marmot is well kept, and as comfortable as possible.

THE GREAT HORNED OWL.

FAR away in the deep forests of North America the traveller has sometimes paused to rest, and to cook his supper, gipsy fashion, on a stick, and by a fire lighted of dry leaves and twigs.

The fire burns brightly, and throws a ruddy light on the trees around. He and his companions are glad to stretch their wearied limbs, and to partake of the repast. No sound is heard of bird or animal, and scarcely a leaf stirs. But, all at once, close to their ears, there breaks out a succession of unearthly shrieks, mixed with wild laughter. It is as if some person were being strangled in their very presence.

THE GREAT HORNED OWL.

Perhaps a minute after, a large dark object sweeps round the fire, still uttering discordant sounds. He is the great horned owl, that was sitting close by, though the travellers did not see him. The agreeable odour of the food roused him, and he came out to see if he might share the feast.

These deep and boundless forests are the home he loves. His voice renders him an object of dread, there is something so unearthly about it.

And among some of the Indian tribes, the priest often carries a stuffed owl at the top of his head, with large sparkling glass eyes. He considers it an emblem of mystery, yet there is no mystery about the bird. He is simply a bird of prey, and with all the habits of his race. All day he hides himself in some thick bush, and is rarely seen to venture abroad; but when night comes, out he flies, and sweeps round and round in search of food. He has a large strong bill, and two tufts of feathers stand up like horns on either side of his head. He utters hideous noises as he flies—so hideous, that you would suppose all living creatures would fly from the sound.

Rabbits, squirrels, mice, rats, and little birds are his favourite food; and if he can rob the hen roost, so much the better.

One that was kept in a cage became rather troublesome, not by devouring chickens, but by barking like a dog.

The master of the house could not sleep for the constant yelp, yelp, of the supposed cur. And a large Newfoundland dog was so deceived that he kept barking in reply. At last the gentleman got up, and took his stick, and sallied forth to drive away the intruder. But no dog was to be seen, and he at last traced the yelping sound to the cage of the owl.

On a clear night, the owl will bark in this way till the morning.

The owls live all the year round in the forests. They make a nest in the fork of some great tree. It is built of a great many sticks piled altogether, and lined with dry leaves and feathers. Sometimes a hollow tree is chosen. There are four pure white eggs laid in the nest, nearly as large as hens' eggs. After the young ones have left the nest, the bones and feathers of many little birds have been found in it. Even the poor woodpecker has fallen a victim.

An owl of this species is now and then—but very seldom—seen in Britain. He is like the bird in the picture, and is called by many names, such as the duke owl, or the great-eared owl. He is said also to live in the islands of Orkney, in Scotland, and to be caught napping now and then on the ledges of the hills.

Nothing is known of his habits here, where he is so rarely seen. But he is not so uncommon in Sweden, and a pretty anecdote is told about him.

There was an owl's nest near to a gentleman's house, and one day the servants caught a young bird that was unable to fly, and put it in a hen-coop. The next day a dead partridge was found lying close by the coop, that had been brought in the night. The next morning some other little animal or bird was found. And this went on for a fortnight. The gentleman and his servants both watched to see who had provided for the wants of the little captive, and there was no doubt but the parent birds had done it.

THE SWALLOW.

IN summer the whole air is filled with flies and insects, myriads of them too small for our eyes to see. The swallow is the bird that from morning till night is occupied in catching them.

She comes as soon as they do. With the early spring the great temple of Nature opens, so to say, her gates and doors, and forth rush living creatures by millions. Countless eggs are hatched, thousands and tens of thousands of tiny grubs and caterpillars spring into life, and in their turn become perfect insects. The chrysalis, that has lain all winter swathed up in some secret spot, now has its bands unloosed, and flutters out on wings.

By the brook, in the meadows, and among the trees, there is a hum and a stir of life everywhere!

The swallow knows that her banquet is spread. When it grew late in the autumn there were no flies, and there could be no swallows. So the swallows, who had been for some time gathering together in flocks or companies, went away, no one knew with any exactness whither.

This part of the swallow's history has to be made clear to us some day. We are told the little bands are seen sailing far over the sea, towards the hot regions of Africa, or other parts of the torrid zone. And they have sometimes rested on the mast of a ship, just, as it were, to take breath. They can go many hours without resting, as we know, for they keep on the wing from morning till night.

Sometimes in winter, in frost and snow, a poor half-dead swallow has been found in an odd out-of-way place. Why she was left behind, we have no means of guessing. After all, the birds keep many secrets from us.

But hail to the spring! The swallows are come! The first that comes is our old friend the chimney swallow, that appears early in April, before the cold weather is over, and is a proof of the old adage, that "one swallow does not make a summer." It is often so much the worse for him. The cold east wind that troubles us so long in the spring kills or benumbs the insects, and the poor swallow cannot find a single fly to eat. Then all that the swallows can do, is to seek out some sheltered nook, in which to drag on a kind of existence on the few insects that, like themselves, have taken refuge there.

One spring the swallows came as usual, but the weather changed, and they went away, and everybody wondered what had become of them. The weather was extremely cold, and no insects were about. At last some one happened to find his way to a sheltered place, in a valley near to the sea, and lo and behold! there were the swallows flying about by hundreds. There they had found a few insects on which to subsist. But they had no great amount of food, and were in a very feeble state, and obliged to settle and rest themselves every few minutes.

The swallows have been seen crossing the Channel, on their way to England, either singly or in flocks. They seem very much tired, and are glad to settle on the boats that are on the water. Sometimes a swallow is so exhausted she can hardly fly from one end of the boat to the other, when a sailor tries to catch her. And a sailor declares that he saw one drop on the sea with her wings expanded, as if quite spent; then, after a few minutes of rest, she rose again refreshed, and flew off merrily.

All the strength of the swallow is in her wing; her feet are feeble, and her legs very small. She could not walk very well upon them; but then she is not much used to walking.

Though her feet are feeble, the toes are strong, and the joints very loose. There is a horny claw at the end of the toe, and she can cling if she cannot walk. She often wants to cling when she is building her nest, and sometimes to hold on to the wall—a feat that seems rather difficult to perform.

But the loose toes are drawn up by some muscles that go up the leg, and then they can clutch quite tight to anything. And she often plasters her nest as she clings, and rests a little on her stiff tail.

Her tail is forked and like a rudder, that guides her through the air. You have a good view of it in the picture, for it is quite expanded, and shows the two white spots beneath.

Her eyes project, and are no doubt very terrible to the insect world. She catches her prey as she flies, with her wide gaping mouth. Her throat is often full of clusters of flies that she has swallowed, and that stick together in little lumps. Some of these are no doubt intended for the little brood at home. There is a sticky substance in her mouth, that helps to catch the flies.

SWALLOWS.

The swallow has been called a mason bird, because her nest is usually made of moist earth or clay. But she always makes a comfortable bed of feathers for the little ones to rest upon.

There are five kinds of swallows that come to us in the summer. There is the chimney swallow, the swift, that makes a squealing noise as he flies, the house martin, about which we shall have something to say, and the sand martin, that makes deep burrows in a cliff or in a sand-bank, and places her nest at the bottom, a little after the fashion of the kingfisher.

In fine dry weather, you see the swallows fly high up in the air, but in rainy seasons they come lower, and skim near the ground. Their way of flying high or low depends upon the insects, for when it rains or is damp, the insects come near the ground; and these are what the swallow is looking for.

For a short time after their arrival, the swallows seem to be doing nothing but enjoying themselves, and recovering from the fatigues of the journey. By-and-by, however, they begin to think of building their nests.

The old birds go back to those already made, and find the old homes out by instinct.

The chimney swallow does not always build in the stack of chimneys, but chooses often to place her nest under the eaves of an outhouse, or even among the rafters of a barn.

One spring a pair of swallows were resolved to build in the rafters of a summer-house. They were not in the least disturbed by the constant presence of persons going in and out, but completed their task, carrying pellets of mud and soft earth in their beaks, and bents of grass to mix with it.

When the saucer-shaped nest was finished, the eggs were laid and hatched without any misfortune happening. The young nestlings had now to be fed every few minutes from morning till night. It was a matter of curiosity how the old swallows would like to pass in and out over the heads of persons sitting in the summer-house, and who, by putting out their hands, could touch the nest. But this fact did not deter them from performing their parental duties, and the little birds grew and throve merrily.

Their heads were soon seen peering above the edge of the nest. As a rule, they were silent; but long before any one could discern the parent bird, they had spied her out in the far distance. They would set up a chirp of delight, raise themselves in the nest, and a row of little beaks would open wide. A minute after, in would dart the mother swallow, without the least fear or hesitation, cling to the rafter by the side of the nest, and pop a fly into the mouth of the one nearest to her, uttering at the same time a peculiar and rather shrill note. As for the presence of two or three persons, she seemed not to mind it in the least.

Her partner was by no means so brave. He would come with a little white moth in his beak, for these moths were plentiful in the meadows, and seemed a staple article of food; but he would turn round and dart away, as if too shy to face the company; nor did he, until after several attempts, venture to bring in the dainty morsel. At each unsuccessful attempt the little swallows raised themselves up and chirped, but only to be disappointed. If the pet dog of the family chanced to be near the spot, both swallows would go into a violent passion. They would fly over his head, and utter shrill cries, and peck at him, without, however, venturing to touch him; all which attacks were lost upon the dog for some time; but at last, as if wondering what the clamour was about, he would raise his head to look, and quietly walk away.

We have not quite finished with the history of these little swallows.

One morning, on entering the summer-house, one of the young brood was seen upon the ground, as though it had fallen or been pushed out of the nest. The mother bird was flying backwards and forwards as usual to feed her little ones, but without taking any notice.

This want of natural affection was set down to the fact that the swallow rarely looks on the ground, and might not see it.

But mark the difference between reason and mere instinct. The little bird was picked up, warmed and revived, and put again into the nest, in the hope that the mother would feed and cherish it. But no. For some cause or other she refused to do so. It might be a sickly bird, or she had more than enough to provide for. At any rate, every time it was put into the nest she flung it out again, until at last the poor little thing died.

The swallow, like most of the birds, will sometimes take a fancy to build in a very odd place.

A swallow's nest was once made in the half-open drawer of a table in a garret. The garret was never used, and the birds flew in and out through a broken pane in the window. And a still more curious place to choose, was the body and wing of an owl that had been nailed against a barn. The nest is in existence now, and has been kept as a curiosity by the family.

The swallows were also known to build in the chimney of a steam-engine that was at work on a farm in Scotland. They reared their young season after season in this, as we should think, rather uncomfortable place, and took no notice of the rattle of the engine, that was almost deafening. They were very timid, although no one ever molested them. If a bird of prey or a cat appeared, they sounded the note of alarm, and darted about in a frantic manner until they had driven the enemy away.

THE MARTIN.

A FEW days after the swallow has arrived, there comes the martin.

They had really started together, and even crossed the Mediterranean in company; but the wing of the martin is smaller and more feeble than that of the swallow, and she lagged somewhat behind.

She knows her way to the nest she left last autumn under the eaves of the house. For she loves the dwellings of man, and is called the house martin on that very account. She is a first-rate builder, and can fix her nest against the smooth surface of the glass itself.

Every one knows the shape of the nest. She lays a foundation of mud, and a layer of soft earth or clay is added day by day, and allowed to become dry and hard before the work is carried on.

This is the outside wall, and is of a circular form. Both birds work hard at the nest, and they seem to have the power of moistening it with their saliva, which makes the substance hard and like glue. Inside, it is lined with hay and feathers, and made quite soft within.

There is an opening in the side of the nest for the old birds to go in and out; and after a time, the little heads of the brood are thrust out to receive the food that they bring. And you may see the mother bird hanging to the outside of the nest by her claws. As soon as one brood has flown, she begins to lay more eggs, and to rear another; and even quite late in the autumn she is still busy with family cares. Indeed, she sometimes has not finished her duties in time, and then she must fly away to a warmer country, and the poor little birds are left behind. This has been known to happen many times. One spring a pair of house martins came back to their old nest, and were seen to draw out the dead bodies of three little nestlings, that had not been ready to fly.

THE MARTIN.

Another pair of birds came to a nest close by, and tried to do the same. But the little ones were too large and heavy, and they could not, with all their efforts, get them out. So they gave up the idea of using the nest for a home, and made it instead into a tomb, by sealing up the hole with clay.

THE MARTIN.

The martins are a very numerous family, and are found almost everywhere. Their flight is not quite so swift as that of the swallow, but the two birds fly about together, and their habits are just alike.

For some days after the young martins have left the nest, they fly about and are fed by their parents. And sometimes they do not seem inclined to leave the nest, and the old birds entice them, or even push them out. And a young bird may be seen clinging to the upper part of a window by its feet and tail, and the parents feed it in this position.

It is amusing to watch the martin building her nest. She plants her tail against the wall, and then deposits the mud she has brought in her bill, like a skilful workman, allowing it to drop into the crevices of yesterday's work. Sometimes she stops to retouch the whole, and make it look smoother. Every time she alights on the nest, she makes a twittering sound; and if the weather is hot, she will now and then take a splash in the pool to refresh herself. She has to provide herself with food, and that takes up a little time; and often she has to wait until the clay gets dry, before she puts on any more.

In bad weather you see nothing of her, for she never attempts to go on with her work. But when the rain is over, out she comes again, and proceeds with the building.

The parent birds will feed their young ones as many as twenty-two times in half an hour. Sometimes they cling to the entrance of the nest, and sometimes they go in. The young one whose turn it is to be fed often sits near the entrance to be ready, and if a morsel remains on its bill, the others snap it up. After a time the parent birds enter on a new series of duties. The young ones are fat and strong, and able to leave the nest, but they are rather afraid. They do not quite know the power of their own wings, and how safely they will be borne up by them. They open and shut them for a long time, as if wishing but fearing to launch on that wide open region in which their parents dart and skim.

One little bird, bolder than the rest, summons all its courage, and springs into the air. The parents welcome it with delight, and all day they sport about among the tree-tops, taking short flights, and feeding and tending the young ones with the utmost affection.

Once a very pretty sight was witnessed, and one that reminds us of a mother coaxing her child to walk by holding out a sweetmeat.

The old bird came to the nest with a fly in his bill, and held it at a little distance, as much as to say, "If you mean to have it, you must fly."

The little martins tried to come out, but did not succeed very cleverly, and nearly tumbled over. A minute after, the mother bird came back with a fly in her bill, and tried the same experiment. But the little ones were afraid, and would not come. She tried for some time to coax them, but at last grew angry, and seemed to scold them for being such cowards. She even took hold of one little bird with her claw, and tried to pull it out, but it clung to the nest with all its might, and she was obliged to give it up.

THE KINGFISHER.

ONCE upon a time, as the story-books say, a house was built in a field that had a steep bank in it, and in the bank a gravel-pit.

When the house was finished, the field was laid out as a garden, and an arbour made in the bank, close by the gravel-pit. People who sat in the arbour on a summer's day looked over a winding river bordered with willow trees, and fringed with reeds and rushes, and covered with the yellow water-lilies, arrow-heads, and many other flowers. And they could see the little willow wren hopping about close by her slender nest upon the reeds, and the water-hen leading out her brood of young ones to enjoy themselves on the stream.

There were very many feathered friends close at hand, but the best of all was to come.

Very early in the spring a great round hole was seen in the bank close by the arbour, and from the state of the arbour itself it was evident that some birds had been at work; but they were too shy ever to let themselves be seen, and the whole affair, both as to the hole and the birds, remained a mystery.

But one day, much later in the spring, a number of delicate white egg-shells lay just below the hole; and scarcely had the discovery been made, when a rushing noise was heard, and a large bird flew out of the hole, and darted down the river, the sun glancing on his plumage of green and gold. No other bird in England is dressed so gaily; it could be but one—the kingfisher.

We must stop a moment to describe him, as he is the hero of our little story.

His body is not elegant in shape, for it is stout and thick, with a short neck and a very long bill. His tail is short, and—if we may use the word—stumpy. His feet are small and feeble; the first toe shorter than the second, and the third longer than the fourth.

The bill is the weapon he uses to catch his prey, and we shall talk about it presently. It is, as you see, much longer than the head, and is straight and rather slender, with a pointed tip.

But his plumage is the most beautiful part of him, and what causes him to be so much admired. The colours are very soft and blended, and yet there is a wonderful brilliance about them. The upper part of the head is a dull green, and each feather has a bar of light greenish blue near the end, that gives it a metallic look. The neck, the sides of the back, and the wings, are of the same green, tinged with purple; but the middle of the back is of a lovely glossy blue, that shines and glistens in the sun; while the tail is of a duller hue.

There is a band of yellowish red from the nostril to the eye, and behind the eye is another band of the same colour. The throat is a yellowish white, and the breast and lower part of the body are of the same yellow red, but of a richer tint.

Thus gaily clad, the kingfisher, in a country where the birds dress very soberly, shines as a conspicuous object.

In his every-day life nothing can be more useful to him than his bill, since he can catch fish with it for the benefit of his family.

He is a famous fisher, and this is why he had made his nest and come to live near the arbour by the pretty winding river. For on a clear still day it was easy to see, not only the reflection of the clouds as they went sailing over the sky, but to catch sight of a shoal of fishes gliding merrily about just beneath the surface.

And in the hot drowsy noontide, when all was still, there would be heard distinctly the plash of the fish as it leaped up to catch a fly, or to breathe the air.

All these facts must have been well known to the kingfisher.

The only difficulty was his extreme shyness and his dislike to be seen. Yet he had chosen the most public place he could find. The arbour was a mere shed supported by wooden posts, and quite open on the side nearest to his hole. The hole was not more than a yard or two off. People sat and looked at it constantly. Not that anything could be seen, for the hole ran in a slanting direction, and had a hollow place scooped out at the bottom, in which was the nest. But the loud chirping noise of the little kingfishers was heard very plainly indeed when the parents were away.

THE KINGFISHER.

The kingfisher had several fishing places. Sometimes he perched on the branch of a willow that overhung the stream, where he would sit for many minutes, lazily resting himself.

He grasped the stem with his small red feet, his glossy back shining in the sun, and his ruby breast reflected in the water below. His long bill was pointed downwards, and his eye intent on watching the tiny fish that sported beneath.

Presently a fish came into the right position, and he opened his wings a little way, and darted downwards with the rapidity of lightning, and, as it seemed, headlong into the water. There was a splash, and in another second he appeared with a fish in his mouth, struggling and twisting itself about.

He struck it against the bough and killed it, then, tossing up his head, swallowed it, and was again on the watch as intently as ever.

The poor kingfisher suffers very much in cold weather, but even in the depth of the winter he is now and then seen plying his trade on the river. One sharp frost the river was frozen except just in the middle, and here a kingfisher was seen hovering over the open spot, as if looking for fish. Presently he dived and brought one up in his bill.

A gentleman who was passing fired his gun, but, we are happy to say, without effect. The bird darted away with the swiftness of an arrow, and was gone.

THE HOOPOE.

There is a family of birds called bee-eaters, that feed upon bees as the swallow does upon flies. They do not live in England, but pay summer visits to the warm countries in the south of Europe. They wear a rich costume of yellow and orange, and have a long beak a little like the kingfisher's, and a body about the size of a swallow.

And they lead the life of the swallow, darting about in flocks on the mountain sides, where bees hum over the flowers of the thyme, and they catch them by hundreds. They have a curious note, that can be heard a long way off, and might be mistaken for a man whistling.

The hoopoe, with its beautiful crest, is a distant relation of the bee-eater. He can raise his crest or put it down, just as he pleases; and he wears a suit of the gayest colours.

He has three toes in front and one behind, and the one behind is long and has a straight claw.

THE HOOPOE.

His native home is in Africa and in Asia, and he is related not only to the bee-eater, but to the humming-bird and the sun-bird. Very rarely, indeed, does he pay us a visit in England, but such an event now and then happens.

THE HOOPOE.

The hoopoe, when he is kindly treated, behaves extremely well in captivity, and even attaches himself to those about him.

We can tell you a little story about it.

Two young birds were taken out of a hoopoe's nest, and brought up in the house. They grew very fond of their owner, and used to follow him about. When they heard him coming they used to utter a joyful kind of chirp, and even fly up and settle on him.

He often brought them a pan of milk, and would let them feast upon the cream, which they seemed to like very much. And then they would perch on his shoulder, and make a great fuss with him. But if he grew tired, he had only to say a word, and they flew away. The room they were kept in had a stove in it, and they used to run behind the stove.

But what was very funny, they always looked at him, to see what sort of humour he was in, before they ventured on their play.

They were very fond of beetles, and had a curious way of eating them. They first killed them, and then beat them with their beaks into a long-shaped ball. This they tossed up into the air, and caught again with their beaks.

One day their owner took them into a field that they might catch a few insects for themselves. They enjoyed this kind of sport very much indeed, but their pleasure was a little spoilt by their timidity.

They seemed haunted by the dread lest some bird of prey should be hovering near; and if so much as a pigeon came in sight they were seized with a panic. They squatted down to the ground in a very curious attitude; their wings were stretched out and their heads leaned back with the beak pointing upwards. They looked more like pieces of old rag that had been thrown on the ground than anything else. The moment the pigeon was gone, up they jumped, and uttered cries of delight.

In his native state the hoopoe is fond of hunting about for insects on some piece of moist ground.

He makes his nest in the hole of some decayed tree. It is made of dry grass, lined with wool or feathers, and has a very unpleasant smell, from the remains of insects that lie about in it.

If the birds cannot find a hollow tree that will suit their purpose, they choose a crevice in the rock or in some old building. The mother bird lays four or five eggs, of a greyish white, spotted with grey or brown.

THE HUMMING-BIRD.

The humming-bird may be called the fly-catcher of the tropics, for he is quite as expert as the swallow.

THE GIANT HUMMING-BIRD.

But his habits and his mode of proceeding are rather different. In the first place, he is, as you know, the smallest of all the birds, and in some cases no larger than a humble bee. But his tiny body is adorned with rainbow colours, and the feathers on his throat have a look like velvet, and change colour every moment in a wonderful and beautiful manner.

There is one great humming-bird that is a giant among his fellows, and the least attractive of any of them. You see him in the picture. He is as large as a sparrow, and is dressed in a rather sober costume of brownish green, the under part of the body tinted with red. The tail is golden green; and the feathers on the throat are velvety, but lack brightness of colour.

The life of the humming-bird is spent among the brightest scenes of nature. He flashes from flower to flower with the rapidity of lightning, and hovers over first one and then another. His food consists of the tiny insects that live amid the petals, and Nature has provided him with a long and slender bill on purpose to pick them out.

Sometimes the bill is straight, and sometimes it is curved. In some species it curves upwards, and the bird is called the avocet humming-bird; in others it is bent like a sickle.

The tongue is slender, and can be put out to a great length. It is made up of two parts, like round tubes, that are laid side by side for nearly their whole length, and then are separate. It is very sticky, and when it is thrust into the flower the insects stick to it and cannot get away. And the edge of the tongue is fringed with tiny bristles, that still further help to entrap them.

The power of wing in the humming-bird is very great indeed. The wings themselves are very long, and the muscles that move them are exceedingly strong. Small as the humming-bird is, he can dart away or continue to fly for a longer time than you would believe. And he can poise himself in the air as the insects do, and keep up a kind of quick vibration with his wings. His movements are so quick that his lovely wings can hardly be perceived unless the sun flashes upon them. In the tropical garden, full as it is of all kinds of splendid flowers, he is seen in perfection, and there are crowds of brilliant humming-birds darting about in gold and purple and crimson.

They are extremely passionate, and if the flower disappoints them in the quantity of insects or nectar it contains, they will tear it to pieces. And they are very brave, and will fight desperately in defence of their nest and their young ones.

Their nest is the most exquisite little fabric you can imagine. It is so small and fairy-like that it is often hung to a leaf, or to the end of a twig.

Often it is made of thistle-down, or of lichens, with cotton and wool inside. Two tiny white eggs are laid in it, which by-and-by become little

THE SICKLE-BILLED HUMMING-BIRD.

humming-birds. At first they are without any feathers, and are more like bluebottle flies than anything else.

There is a huge spider in that part of the world as big or bigger than the humming-bird himself.

He makes a great strong net in which the brilliant little creature is often caught. And what is almost worse, he comes with his hideous long legs

THE TOPAZ-THROATED HUMMING-BIRD.

running up the tree, and gets into the nest. Then he kills both the parents and the young ones.

The humming-bird family is a very large one, and contains all manner of brilliant little creatures, with names such as the topaz, and the amethyst, and the ruby-throated humming-bird. And sometimes the tiny gem is adorned with an elegant ruff on its neck, that it can set up and down at pleasure, and then it is called a "coquette."

There are some humming-birds that live in the forest, and keep in the shadow of the trees. This is very different from the usual habit of the family, as they are creatures of the sun and cannot live without it.

A little humming-bird was once put into a shady room, where the sun did not shine. It became cold and benumbed, and dropped down as if it were dead. The people of the house had to carry it out of doors, and lay it in the full glare of the sun. Then it soon began to revive, and became as lively as ever.

But the humming-bird of the forest does not seem to care about the sun. It searches for insects among the leaves of the trees, and threads its way along from branch to branch.

This curious humming-bird not only lives in the shade, but does not venture out until evening. It has almost the habits of the owl.

People cannot rear the humming-birds in England, the climate is much too cold.

A young man once saw a humming-bird sitting on her nest, and he contrived to cut off the branch, and carry away bird and nest together. The ship was just ready to sail for England, and he took it with him on board.

The mother bird soon hatched her eggs, and grew very tame, and would eat out of his hand. But she died before the ship reached England.

He contrived to rear the two young birds, and they lived some little time after they reached this country. A lady used to feed them out of her mouth, but in spite of every care the poor little things fell victims to the climate.

THE SUN-BIRD.

THE sun-bird is, as you see, a near relative of the humming-bird, and wears the same gorgeous attire. He has been called the humming-bird of Africa, because that part of the world is his home, and there a vast number of sun-birds are found.

The family name of *Nectarinidæ* has been given because the bird sips the nectar of the flowers, and was once supposed to live entirely on honey. But it is now well known that the sun-bird feeds very much on insects, and only gives zest to his banquets by a sip of the juices of the flowers.

The name of sun-bird is very appropriate, since the sunlight has a remarkable effect on his plumage, and makes it look like gold and precious stones. This effect is really produced by the bird, as he moves his muscles and changes the position of his feathers, throwing them every moment into a different light.

The upper part of the head and throat are blue, changing into violet; then, at the back of the neck, there is a patch of red; and the wings are purple and green, and the breast is yellow.

The tail is of scarlet, and the feathers reach some distance.

THE COMMON TREE-CREEPER.

If on the occasion of some summer holiday you go into the woods, and look well about you, you are almost sure to see the little tree-creeper.

The best way is to sit down on the stump of some fallen tree, and quietly to watch.

By-and-by you will see a very small brown bird, the smallest we have in England, except, perhaps, the dear little wren. He has a short neck, and a long, rather curved bill, and a stiff little tail.

His foot has three toes in front, and one long one behind, and are all armed with rather long claws. The joints roll about loosely, in the same way as those of the swallow. But as for his daily habits, they are more like those of the woodpecker. He and his family are very useful in clearing trees of insects that the swallow cannot get at. They lie snugly hidden under the bark or in cracks, but the bill of the little creeper is sure to find them; and though his tongue is not quite like that of the woodpecker, yet he thrusts it

SUN-BIRDS.

out in the same manner. Its tip is hard and sharp, and can run through the insect like a spear.

In the picture you see him at his daily work in the woods. He is clinging with his sharp little claws to the bark of the tree. He has mounted up by a number of short jerks, and every time he jerked he uttered a shrill small cry.

THE COMMON TREE-CREEPER.

You would be amused if you sat and watched how he climbs and winds about. Now he goes right round the trunk, and you lose sight of him. But not for long; in a few minutes he comes back. He crouches close to the tree, and presses his tail against it. You see him pick something out of a crack with his tongue, and give another little jerk forward. He is never at rest for a moment.

His plan is to begin near the root of the tree, and direct his course upward.

And we must tell you that he is far more nimble than the woodpecker, and runs along the lower side of the branches with the utmost ease.

He is very cunning; and if he should chance to see you, he will try to keep on the opposite side of the tree, so that you can only catch now and then a glimpse of him. All this climbing and threading seems mere play to him. When he has wound along the branches till he reaches the top of the tree, and can go no farther, he flies down like an arrow, and alights at the root of another tree. Then he begins again his upward course, and climbs and threads with as much zeal as ever.

His note is a little low "cheep, cheep," that he utters very often.

Early in spring the tree-creeper builds his nest. He chooses some hole in a tree where a branch has been broken off, or where the woodpecker has made the hole before him, or even in a hole in the wall.

The nest is made of the usual materials—moss, grass, roots, and feathers. These are put into the hole in great quantities, so as to make a firm foundation. The mother bird lays six or eight eggs, of an ash colour, marked with dark red spots. She sits very closely, and will hardly move even when any one comes near her. Her partner is very attentive, and feeds her constantly during this anxious period.

THE NIGHTINGALE.

IN the middle of April, in certain favoured spots in our island, there is heard, both day and night, the voice of the nightingale.

He is the finest singer—the most perfect musician—we have in our woods and groves. Who does not love to catch his thrilling notes! Who does not stop and listen with delight!

The nest of the nightingale is slightly made—so slightly that it is not easy to take it away. It would fall to pieces unless a string were tied round it. And it is placed in a lowly position on the ground, at the bottom of a hedge, or even in some small hollow place. When the young nightingales are hatched, the parent birds feed them with green caterpillars and other insects. All the time they are being hatched the beautiful songs are heard, but when the little birds have to be fed the melody ceases. Then nothing is heard from the once brilliant performer but a low croak, and a snapping noise if he is alarmed. He is busy helping his partner attend to the little ones.

And in August, quite in the midst of the summer, he leaves us altogether.

We have almost forgotten his appearance, but this is not at all striking. He is a plain brown bird, with rather a rich tint, and the under surface of his body is a dull white. His voice is all we care for.

As a rule, the nightingales are not at all plentiful, except in certain places. About London they sing in a delightful manner, and several of our southern counties are highly favoured by their presence. But the birds seem to dislike the northern part of the country, and are rarely seen there.

THE ROBIN.

Who does not love the robin? He is everybody's friend; and he of all the birds never leaves us.

Ever welcome, he visits our window in the bleak winter's morning to receive the crumbs allotted to him, and to repay us with his brisk little song. In the summer, amid the crowd of feathered friends, he seemed to pass from our notice. And he had his own affairs to look after—to build the nest, and rear the young. Now this busy, anxious season is over, and he bethinks himself of his old friends, and draws near the gardens, where perchance a stray grub may remain or be hidden under the grass, and where the friendly hand will give him crumbs.

His appearance and his attitude are known to every one. He stands, his head a little raised, his wings drooping, his mild, pleasant eye beaming with intelligence. Sometimes he spies a worm wriggling in the grass, and he gives a hop towards it, and pecks it, and devours as much as he can. Then he resumes his former position. Now and then he sings his few sweet notes from a wall or a decaying stump. His flight is rapid, but consists of short quick starts from one place to another. Now he is on the bush, now on the fence.

THE NIGHTINGALE.

If you are digging in your garden, he drops down silently close beside you, so that you can almost touch him; but if you look at him, he hops on to the railing, or to some safe place. The next moment he is down again, looking for worms.

Though sociable with us, he is very quarrelsome with his neighbours, and has a great dislike to the sparrow, and will peck at and fight fiercely with him. And he has many a fray with his brother robins. Two of them were once seen contending with such violence, that they rolled together on the garden walk. Indeed, one would have killed the other if they had not been parted.

Nor is the robin on such friendly terms with man at all seasons. It is hunger that makes him so tame as to venture almost within our doors.

Whether he knows it or not, he is in most cases perfectly safe. Ever since the time of the "babes in the wood," no one would think of hurting a robin.

The nest of the robin is placed under a hedge or bush, and is rarely found in a tree. It is large, and rather loosely made of grass and moss and decayed leaves. And five eggs are laid in it, of a reddish white, faintly marked with purple.

Sometimes the birds choose a very odd place for the nest, and are very resolved to carry their point.

A pair of robins once began to build in a tall myrtle plant that stood in the hall of a gentleman's house. The nest was objected to for many reasons, and pulled down. The birds then began to build in a still more curious place. They chose the cornice in the drawing-room, and began to carry moss and leaves there. Of course, such a proceeding was stopped at once, and the nest removed.

The robins, nothing daunted, now began to build a third time, in a new shoe that was placed on a shelf in the dressing-room. They were allowed to go on with the work until the nest was finished. Then, as the shoe was wanted to be worn with its fellow, the nest was taken carefully out, and placed in an old shoe instead. The robins did not seem to mind the change in the least. They filled up the under part of the shoe with oak leaves, and very soon the eggs were laid and hatched. The windows of the room were kept open a little, so that the parent birds could go backwards and forwards as much as they chose. Nothing could exceed their tameness. They made themselves perfectly at home; and when the gentleman who used the dressing-room was shaving in the morning, the two birds would settle on the top of the glass with worms in their mouths, and look at him with the utmost freedom.

A robin that lived in a garden became so tame that he picked worms

ROBIN RED-BREAST.

from the hand of the gardener, and would sit on his knee at dinner-time to have a share of the crumbs. The gentleman who owned the garden was very fond of the robin; and when winter came, he put him into an empty room, that he might have shelter as well as food. And lest the bird should be lonely, he found a partner for him, to cheer his solitude.

Early in the spring the robins were let out to fly. But they had been very busy meanwhile, and had built a pretty little nest, out of what few materials they could find.

A few withered stalks of horehound hung from the ceiling, to be used as a medicine if any of the family took cold. The birds had pulled it down, and used it for the outside of the nest. The inside was lined with the down of another plant, kept there to feed a tame bullfinch. And the result of all was a very tolerable little nest.

THE REDSTART.

THE REDSTART.

THERE is a small class of birds that are related to the robin and the nightingale. They are known by the name of redstarts, and are rarely met with in England. One species only is at all common among us, and is called white-pointed, because of the white mark over the eye of the bird. He is handsomely dressed, in grey, black, and orange. The throat and back part of the head is black, and the breast and tail light reddish orange.

The redstart has several relations that live on the continent, and rarely, if ever, come to England. He himself is very often seen, and his habits watched by the naturalist.

He does not spend the winter in England, but comes about the end of April. He is very lively in his disposition, though he is small; and is well known by the white patch over his eye, and the manner in which he jerks his tail up and down at all times, either when perched or when on the ground. He catches insects on the wing. The nest is placed in a hole in the wall, or in a hollow tree, or even in the crevice of a rock. It is made of roots and moss, and is well lined with hair and feathers; and there are six or seven eggs, of a light greenish blue. The young redstart does not attain the beautiful plumage of his parent until he has left England on his autumn tour. The hen birds have no white on the forehead, and wear a more sober costume. When the redstart arrives here in the spring, he makes his appearance in the gardens, or near some old wall. He is not often seen, for he is very shy, and keeps close to his home. When the nest is built, and his partner sitting on the eggs, he will perch on a branch close by, and sing his little song, which is very short, and has not much music in it. He sings almost all the day, from the very early morning; but the melody is rarely noticed by any one except his partner.

He does not run or walk, but makes his way by a number of flying leaps. His food is obtained by catching insects on the wing, and sometimes he alights to pick up some worm or grub he has espied.

A pair of redstarts took up their abode in a hole in the gable end of a cottage, on the public road. A weaver lived in the cottage, and from five in the morning until ten at night those noisy looms were working away. Yet the birds never seemed to be annoyed, and reared their young in safety. This is rather curious, for they are very shy birds.

When the redstart is caught, and put in a cage, which sometimes happens, he will sing both night and day, and can be made to imitate the notes of other birds. There was one of the family that used to sing the "Copenhagen Waltz." But the birds never live long in a state of captivity. They cannot have the food most suitable for them, such as the eggs of ants and different kinds of insects; and are subject to fits and many kinds of diseases.

About the end of September the redstarts go away to a warmer climate. They are very abundant on the shores of the Mediterranean; and in France they are caught in great numbers, not as cage birds, but to be eaten. Small as the bird is, his flesh is thought to be a luxury.

F

THE WHITE-THROAT.

A PAIR of white-throats built their nest in a shrubbery, and the kind owner of the place watched over them with the utmost care. The task had been

WHITE-THROATS.

completed, and the mother bird was sitting on her eggs, when, one morning, as the gentleman was walking at some distance from the nest, he saw a broken egg-shell lying on the ground.

"Ah!" thought he. "Some accident has happened to the white-throat. A magpie or a weasel has run away with her eggs."

He went hastily to the nest to look, when, to his joy, it was full of young birds all newly hatched. And he found out that the mother had taken the broken shell, and dropped it far away from her home, lest it should disclose the beloved spot to some passer-by.

The white-throat does not live in England in the winter. It comes with the swallows, and loves the green lanes and hawthorn bushes of the deep country. As soon as it comes it begins to warble a few sweet notes now and then, and to flit about from bush to bush in a restless way. Sometimes it flutters in the air, singing all the time, and then drops down again.

As it sings, it swells out its throat, and puts up the feathers on its head. And though it is a small, slender bird, yet this habit makes it look rather stout. Its plumage is very soft, and of a brown kind of red; and the throat is pure white, from whence it has its name. When it comes in the spring, it has just put on its summer suit; but by the end of the season the colours fade, and the tail feathers get very ragged.

The white-throat has several relations. One is called the garden warbler, and is not quite so familiar. Its song is sweet and mellow, and it warbles as it threads its way round the trunks of trees in the copse, or through the brakes and bushes. It has several notes, and some of them remind one of the blackbird, only that they are sung more hurriedly. It lives in thick hedges, and makes a nest like that of its relative, only it puts a little moss outside. It has also the same love of fruit, and comes to the gardens when the strawberries and the currants are ripe. But though it cannot resist any kind of fruit, it is still a welcome visitor; for it is the only one of the family that will feed on the troublesome caterpillar that lives on the cabbage, and does so much mischief. It is but just to set one fact against the other.

It has been seen darting into the air to catch insects, after the fashion of the fly-catchers. It sometimes takes its stand on some post, or stake that has been put to a dahlia plant, and' watches for its prey. When an insect comes by, it darts with its bill upwards, catches the fly in a second, and goes back to its post.

LONG-TAILED TITS.

THE LONG-TAILED TIT.

VERY often, in a summer ramble, or when walking in the garden, a peculiar scraping note is heard from among the trees. There are two sounds, one like the word "churr! churr!" and then a shrill note like "twit! twit!" You may know at once that the sound comes from the long-tailed tit, the head of the tit tribe, and the handsome bird you see in the picture.

There are many of the tit family on every hand. They are birds of a small size, and do not exceed a sparrow in bulk. They have a short straight bill, with rather a sharp point, and they use it to feed on various kinds of seeds, and even to split open a cone. But their favourite food is insects, and they destroy millions that are out of the way of the swallow.

They do not refuse scraps of meat or any morsels thrown out from the kitchen door; and on a winter's day, when the little birds come round our houses to ask our bounty, some of the tit family are very likely to be present. They give little rapid flights, and utter shrill cries when picking up their food. Small as they are, they have a great deal of courage, and will attack birds much larger than themselves.

The long-tailed tit is known from the rest of his tribe by the long handsome tail that Nature has given him. His plumage is very soft and thick. His head, throat, and breast are white, and there is a broad black band over the eye.

His thick tufted plumage makes him look as if he were muffled up to the chin, and that is why some people call him the "mufflin."

In winter the long-tailed tits fly about in troops, and follow each other in a file, their long tails streaming behind them.

They are very social birds, so much so that, in the season when the mother birds are sitting on their eggs, their partners will feed together in the most friendly manner.

The long-tailed tit is famous for making one of the most beautiful nests that is known. The other tits place their nests in a hole, but the nest of which I am speaking is fastened to the twigs or branches of a bush not many yards above the ground.

THE WAGTAIL.

There is not a more lively bird anywhere than our friend the wagtail. It is true he does not entertain us with a delightful song, like some of our feathered friends, but he amuses us none the less.

He is very tame and sociable, and rather likes to be in our company. And he is handsome to look at in his glossy plumage of grey, black, and

WAGTAILS.

white, all blended together; and he has a brisk way of walking, or rather skimming about with very light footsteps, as if he trod on air.

He is never so happy as when he can get to the side of a brook or stream. He will wade about in a shallow place, looking for insects and worms, and holding up his handsome tail to keep it from the wet. Sometimes

THE WHITE WAGTAIL.

he perches on a stone, and stands jerking and twittering, or he ventures quite into the water as far as it will be safe, to see what he can find. His footsteps are so light they leave no impression on the mud, and he runs over it without sinking in. He and his family are great friends to the cattle when they stand grazing in the fields. On hot sultry days the flies are very troublesome, and the cow has to keep switching her tail to drive them off. But presently there is sure to come a little party of wagtails in search of these very flies. They run, and jerk, and dart, and spring, and at every movement there is a gnat or a fly the less.

The nest is sometimes placed in a snug place among the grass, or on a heap of stones, or even in a hole in an old wall.

It is made of leaves and stems of grass, mixed with moss, and lined with wool and hair. It is rather large and rough looking, and all kinds of hair are used to line it.

The mother bird lays five or six eggs on this warm bed : they are of a greyish white colour, and spotted with grey and brown. When the little wagtails are hatched, their parents rear them with the utmost care, and even when they have left the nest will help them to get their own living. If the mother bird finds an insect larger than usual, she will not eat it herself, but give it to her young one.

We must tell you that the cuckoo sometimes drops her egg into the wagtail's nest. Then the mother wagtail hatches it as if it had been her own. But the young cuckoo grows so very large, that its foster mother has to settle on its back in order to feed it properly. Then the cuckoo turns its beak round for the food, as if it understood the whole process.

The wagtail in the picture is called the mountain wagtail, and lives in the mountains of Europe. It is a deep grey colour on the back, and the under parts are yellow. The throat is black, and there is a white line between it and the back.

The mountain wagtails are met with in all mountainous countries.

In the winter they migrate to the tropics, like their relations the friendly little wagtails we have been speaking about. They are very sociable, and run about near the houses in the same quick and nimble manner as with us, and have the same twittering note.

THE WREN.

THE appearance of the wren is very peculiar. The little body is short and full, the bill very slender, and the tail is generally erect. It jerks its whole body as it hops about, uttering a note that sounds a little like "chit! chit!"

THE WREN.

It is always cheerful, and never in the most gloomy weather seems out of spirits. When the sparrows and the finches sit dolefully on the twigs, with their plumage wet and ruffled, the wren is as compact and as neat as ever, and hops about as if the weather made but little difference to it. When the great drops of rain are falling in torrents, its bright eye will peep from a hole, or its knowing little head pop from behind a wall, as merrily as possible.

The nest is such a wonderful piece of workmanship, that people have never been weary of praising it. The clever little bird takes the utmost pains in its construction. All kinds of plants enter into its composition.

Here are mosses to make the walls, and it is arched over with fern leaves, and stems of grass, and twigs of trees, all twisted and matted together in the most beautiful manner.

The nest is in the shape of a ball, and has an opening in the form of a low arch, just large enough to admit the wren, and the inside is lined with soft feathers to make a bed for the young ones. Sometimes the nest is placed in a hole in the wall or of a tree, or amongst the thick boughs of the honeysuckle or clematis. Indeed, it is put in all kinds of places.

The mother bird is the real architect, and her mate sits by and sings. He does not give her much help, thinking, perhaps, that his music is enough. The song is rich and mellow, and can always be listened to with pleasure.

While he sings his partner works. She fetches and carries the moss and the leaves, and is often seen dragging a bundle almost as large as herself.

If any one looks at her, she disappears into some nook or corner; but we may be sure her bright eye is peering from the retreat, and eagerly watching the movements of the intruder. The instant he is gone, out she comes with her usual bob and jerk, and contrives to drag the bundle onwards to the nest with unabated zeal.

The only tools she ever uses are her beak and claws, and she gives the nest its shape by turning her body round and round, and pressing against the sides with her breast and wings. When it is finished, the texture is so close that not a drop of rain can enter, and in this snug little dwelling the mother wren lays a great number of eggs.

Very soon begins the task of feeding the young birds, a task which the little wren carries on with the same untiring energy. The little mouths are always open, and it is no easy matter to fill them. All day long the careful mother goes backwards and forwards, bringing flies and insects, and whatever food she can find, to satisfy the craving of her brood. She has been seen to go to and fro as many as two hundred and seventy-eight times in a day!

The little birds, with such good feeding, grow very fast, and at length the happy moment comes when they may leave the nest.

By this time they are getting very tired of it, pretty and snug as it is, and their little wings are impatient to flutter in a large space. They want to hop and jerk like their parents, and show the same lively disposition, though their plumage is at present rather different.

At first they run about on the ground, and hop among the bushes, attended by their parents, who do not consider their education quite complete. If any person or animal approaches, the mother wren becomes very much alarmed, and utters a loud "chit! chit!" The little birds hide themselves in any hole or corner that is near, and the whole family disappears as if by magic.

In spite of all the pains and labour taken by the wren to build her nest, the beautiful little fabric is often unused. So many of these deserted nests have been found, that the matter has been inquired into by the naturalist, and a rather curious fact has come to light.

It is thought by some persons, that while the mother wren is sitting on her eggs, her partner occupies himself with making a number of nests, one after the other. But he does not line them in the same delicate way, nor does he conceal them with the same dexterity. They are often found by the country people, and go by the familiar name of "cock nests."

However, one thing is pretty certain, that during the severe weather the wrens and their little ones take refuge in these spare houses, and lie there snug and warm.

THE GOLDEN-CRESTED WREN.

PERHAPS few of the perching birds construct a more beautiful nest than the little golden-crested wren.

He is a near relative of the common wren in her plain suit of brown, but he wears a much gayer costume. The crest on his head is bright yellow, tipped with orange, and there is a great deal of yellow about his plumage. Altogether, he is rather a showy little fellow.

He loves the deep woods, especially if he can find a few fir-trees. Here he and his partner like to make the wonderful little dwelling in which they intend to rear their young ones; and here they can run about on the branches and find plenty of insects.

The nest is placed under the branch of the fir-tree, and hangs to it by some little cords, that are in reality the twigs of the fir woven with the moss of the nest.

GOLDEN-CRESTED WRENS.

The dark heavy branch of the fir is a capital screen for this beautiful hanging nest, that is lined with feathers, and has a hole in the side for the birds to go in and out.

The mother bird lays from six to ten eggs, of a pale reddish white. She is a very careful mother, and feeds her young ones every two minutes.

The little golden-crested wren abounds in England, and does not seem to mind the cold of the winter. He is all life and activity, and may be seen hopping and clinging among the branches, and making himself quite happy and contented, let the weather be what it may. He has a sweet voice, but

THE GOLDEN ORIOLE.

not very strong, and it cannot be heard at any distance. And he is very tyrannical in his disposition; and, small as he is, often gets the mastery over the other birds.

THE GOLDEN ORIOLE.

IN the early days of spring, when the leaves are budding forth, many of our feathered friends pay us a visit from other countries. The golden oriole, a beautiful relative of the thrush, comes now and then to spend a little time with us. His visits are few and far between, and he goes away in the autumn to a warmer climate.

In this country he is very rare, and often several years pass without a glimpse of him. But in the sunny land of Italy the orioles abound, and there their domestic life can be better studied.

The oriole is extremely handsome. His beak is an orange brown, and there is a dark-coloured streak from the base of the beak to the eye. His head and neck and body are a bright gamboge yellow, and his wings are black. The mother bird has not the streak under the eye, and is much more plainly dressed.

The oriole has some foreign relations that make long purse-shaped nests; but he and his mate do not follow their example. The beautiful little nest in the picture is, as you see, flat and saucer-shaped, and is fixed very safely in the fork of a branch. It is made of wool and the stalks of grasses, woven together in a most wonderful manner; and four or five smooth white eggs, spotted with brown, are laid in it.

The birds choose a very retired spot for the nest. They fix upon some lonely grove, and make it under the overhanging leaves and branches. They are very brave in defending their young, and would do battle with almost anybody.

THE MOCKING-BIRD.

IT is not in England that you must expect to hear the mocking-bird. It is true that in our woods and thickets one bird will mimic the note of another, but this is in a very inferior way compared to that prince of mimics, the mocking-bird.

THE MOCKING-BIRD.

He is about the size of a thrush, and lives in the warm parts of North America, where the orange-tree is seen in the gardens and the groves, and the fields and the forests are adorned with millions of flowers. Here the magical song of the mocking-bird is heard to perfection. He is the sweetest and the

THE MOCKING-BIRD.

most wonderful of the forest musicians, and his clear strong voice seems to fill the whole air with a flood of harmony.

The mellowness of his song, and the variety and brilliance of its performance, can hardly be described.

He begins to sing early in the season, when he has chosen his partner. He flies lightly round her, his tail expanded and his wings raised, dancing, as it were, with delight. Then he alights on a tree, and seems to pour out all his affection in his song. And more than this, as if his own powers of expression were too few, he begins to imitate the songs of all his neighbours in the grove. In the clear summer night, when the moon shines on the rustling tops of the forest trees, he will sing like the nightingale, and you might take him for one.

But his power of imitation is not always used in such a harmless way. He likes to play certain tricks that are rather mischievous. He will amuse himself by uttering a loud scream like a bird of prey. His feathered neighbours hear it, and are struck with terror: they think the hawk is close at hand, and hide themselves under the boughs. He will keep them in this state of panic just as long as he pleases.

When in this sportive humour, he will torment the hunter who has come out with his gun to shoot some birds.

The mocking-bird sees him from where he is perched on some bough or spray. He begins directly to imitate the notes of most of the birds. The notes are so clear and distinct that the sportsman thinks the songsters are close at hand, and he spends a great deal of time in looking for them. But in reality the birds are a long way off, quite out of reach of the gun.

In the midst of all these frolics and musical performances, the serious business of life has to be thought of, and the nest has to be built.

Then the mocking-bird and his partner set to work and choose the most suitable place for it.

When the choice has been made, the two birds pick up dried twigs, leaves, and grass, and begin to arrange them in the forked branch of a tree. When the nest is finished, five eggs are soon laid in it, and while the mother bird sits upon them he perches close by and sings his sweetest songs. At the same time he keeps a sharp look out for insects. Now and then he drops from the branch, picks up one in his bill, and carries it to his mate.

He is by nature a very brave bird, but at this time nothing can come near the nest without being attacked. Even his natural enemy, the cat, is driven away if she ventures to approach; and as for his battles with the snake, nothing can exceed them in fierceness.

The young birds are hatched in about a fortnight, and there are as many as three broods in the course of the summer. There are not so many insects when the last brood comes out, and they are rather stunted compared to the others. The mocking-bird remains all the year round in some of the warmer states; but in the colder parts he is a summer visitor, and then goes away as the swallows do.

Like the nightingale, he is often caught in a trap, and sold for the sake of his voice. But even when a captive, he is just as lively and as mischievous. He imitates all the sounds in the house, and plays a trick upon everybody. He whistles like the master, and the dog jumps up and thinks it is wanted. Then he makes a pitiful squeak, and the hen clucking outside in the farmyard is in a fright about her chickens, and fancies that one of them is hurt. Then he sings like the canary, or the thrush, or the nightingale, and affords the utmost delight to his listeners. As he sings he appears to be in a kind of ecstasy, and flutters round the cage as if he were dancing to his own music.

The dress worn by the mocking-bird is not very gay, and yet he is on the whole rather handsome. The upper part of the head, neck, and back are of a dark ash colour, and the wings and tail nearly black, the upper feathers tipped with white, in the manner you see in the picture; the lower part of the body is a brownish white, and the legs and feet black and very strong.

The food of the bird consists of berries of the red cedar and the myrtle and the holly, and many others that are found in abundance in the thickets and swamps. And he is very expert in catching winged insects, that abound in the warmer districts even in the winter.

He has a graceful way of moving about on the ground, and keeps opening his wings and closing them as the butterfly does when she is basking in the sun. His flight is performed by short jerks of the body and wings, and each time the tail is twitched strongly. When he is travelling he goes from tree to tree, or at most across a field, and never rises higher than the top of the forest. His common note is rather a mournful one, and he utters it as he flies.

His courage is so great that not even the bird of prey likes to attack him. But there is one kind of hawk that flies very low and very swiftly, and sometimes carries him off with a sudden swoop. But if the hawk misses its aim, the mocking-bird turns upon it with great fury, and calls all his friends to his help. The hawk darts away so fast that none of them can overtake him; but the warning note has been uttered, and the birds are on their guard. The enemy is very likely to go without his prey for that day at least.

THE SONG THRUSH.

THE family of our well-known songster is a very large one indeed, and includes nearly a hundred and twenty species. But our friend in the picture, who is pouring forth his melodious song, is called *musica* on account of his vocal powers.

Who has not heard him some mild day in the early spring, or rather towards the close of the winter? He is seated on the top of a tree that is as yet bare and leafless. The country has no glad look at present; the fields are bare and barren, but the very sound of those clear delicious notes inspire us with joy. "The thrushes," we say, "are beginning to sing."

He is a plain bird, with brown spots on his breast. But if he wore all the colours of the rainbow, he would not be any dearer to us on that account. It is his voice we prize, and we do not look at his dress.

His song is very full, and has many variations. Some persons compare it to that of the nightingale; and, indeed, he may rank next to that prince of song. Early in the morning and late in the evening he is heard, perched often on the same twig.

His domestic habits can easily be studied, for his family abound on all sides of us. His nest is built very early, for he and his partner rear several families of young thrushes in the course of the season. It is rather a rough-looking fabric outside, but within it is very nicely finished, and looks like a smooth hard cup, quite water-tight.

As a rule, the nest is placed in a bush or a hedge.

In Italy the thrush feeds delicately on grapes, for he is a berry-loving bird. But in England, where we have no vineyards, his principal food is of an animal character. He devours slugs and earthworms, and you may see dozens of thrushes on the lawn after rain, walking about and picking up the worms. His bill is as long as his head, and is very useful to him in procuring his food. He has a great fancy for snails, and he breaks the shells to get them out; and he has been seen to place the snail between two pebbles that he used for the purpose, and hammer with his beak until the shell was broken.

Towards the autumn a great many thrushes come from the north of Europe to winter in England. Great numbers of these birds are often met with at that time, and they fly in company with the redwings and the fieldfares. Indeed, they are often sold in the market with other small birds as an article of food; but such diet is not very popular in England.

We must not forget to mention that the thrush is a very orderly bird, and does not allow any litter near the nest. If any stones or rubbish happen to be

THE SONG THRUSH.

there, he carries them away. And the parent birds keep the little ones very clean. A naturalist once rubbed the backs of the young thrushes with mud, to see what the old birds would do. When they came back they were quite in a fuss, and took grass in their bills, and rubbed away at the feathers until there was not a speck of dirt left upon them.

THE BLACKBIRD.

The two most famous singers among our feathered friends are the blackbird and the thrush, and they both belong to the same family. Indeed, the blackbird might be called a black thrush. Yet his note is very different from that of his relative. It is not so lively, but makes up for it by richness and mellowness. On a spring evening, especially if it rains, what music is more delightful than that of the blackbird, as he pours out his heart to his partner?

You may often see the blackbird on the lawn hunting for worms. He keeps his eye on the ground for a few minutes, and then hops up to the spot and gives a great peck. Then he drags out a worm and breaks it in pieces, swallowing it bit by bit. He is more shy than the thrush, but more lively in his disposition. He delights in the side of the hedge or of the wall, for there he can easily procure worms and snails. In the winter, when there are none to be had, he will devour berries of the hawthorn, or even betake himself to the farmyard, and pick up what grain he can find. If he is disturbed, he utters a chuckling sound, a little like the magpie's.

The blackbird does not sing entirely to amuse his mate, for he is often heard in the mild days of winter, when his notes are doubly welcome. In March he thinks about making his little arrangements for the summer, and he and his partner begin to build their nest. They place it in a bush of some kind or other, such as a laurel or a honeysuckle, or even a hole in the wall will be chosen as the favoured spot.

It is a large nest, and the outer part is made of stalks of grass and twigs, woven with moss. Next there is a layer of mud, and then another layer of roots and stalks and leaves. The blackbirds rear two broods in the course of the summer, or even more. A pair of blackbirds were known to build four successive nests during the summer, and to rear seventeen young ones.

There is a pretty story about the domestic life of the blackbird, to show that he is not free from many little cares. He and his mate had a brood of young ones, and kept feeding them all day long. But the mother had brought a large worm, and gave it to one of the brood. Then she flew away, but when she came back there was the worm sticking out of the mouth of the poor little bird in a very uncomfortable manner. The mother blackbird seemed taken quite by surprise, and uttered a cry of distress. Upon that, forth came her mate from a bough close by, as if to see what was the matter.

THE BLACKBIRD.

The two birds did all they could to make the worm go down, but in vain. Something prevented it, and it was a little time before they found out what it was. At last the blackbird perceived that the end of the worm had caught on the feathers of the breast, and was held tight, so that it could not be moved. He at once began to remedy the mischief, and to disengage the worm. It cost him a great deal of trouble to do so, but he succeeded at last, and then he held the worm in his bill, and gently put it down the throat of the little one. The poor little blackbird had narrowly escaped death, and as it was, it lay almost as if it were senseless for a long time. But the parent bird went on to his twig close by the nest, and sang one of his sweetest songs, as if rejoicing that the danger was over.

THE WATER OUSEL.

THERE are many sweet and retired places still left in England. Here and there, in some hilly county, you come on a wild glen, with a clear gushing stream dancing along, often over stones and jutting pieces of rock. Here ferns grow in abundance, and patches of purple heather, and here the birds rejoice in a happy security.

Just such a spot as this is where the merry little dipper dwells. He is a neat compact bird, with rather glossy plumage. His throat is pure white, and his breast a chestnut brown, blending into grey. The white patch on his breast is seen sparkling in the sun as he stands perched on a stone in the stream. Everybody likes to see him. The tourist, as he makes his way through the delightful nook, pauses to say a kindly word to him. The angler knows him well, and so does the shepherd on the lone hillside. The naturalist stops to point out a curious fact that perhaps you do not know. The dipper is a relation of the thrush, and yet he seeks his food in the water. He fishes in a small way on his own account, and it is well if he is content to do so in some retired spot like this.

The owner of some stream, who is trying to preserve his trout and salmon, finds great fault with the dipper. He accuses him of devouring the eggs of these valuable fishes, and persecutes him to the death.

Here, however, are neither trout nor salmon, and we may pause a moment and watch his proceedings.

He is still standing on the stone, his tail jerked up, his wings drooping, and his legs bent. Presently he plunges into the water, in the most fearless manner, and dives under the surface. He does not, however, dive head foremost, like the kingfisher, but takes a gentle walk into the stream, then opens

THE WATER OUSEL.

his wings a little, and goes beneath the surface. When under water he makes the most of his time, and picks up little eels or fishes, or all he can find. And he has been seen to fly under the water after his prey, waving his wings as if he were in the air; then he comes to the surface again as buoyant as a cork.

Sometimes he tumbles about under the water in a very odd way, with his head downwards, as if he were picking up something.

He does not go very far under the water, but soon comes up again and settles on a stone; nor does he walk on the bottom of the stream, for he is more of a hopping than a walking bird. His curved claws are very little use in running, but they can take hold of the slippery stones, and give him a firm footing. He utters a note as he flies like the word "chit;" but, besides this, the dipper can warble a pleasant little song of his own, and that he keeps on singing all the year round.

The nest is placed in a snug place on the bank of the stream, or in the roots of some tree that overhangs the water, or in a retired spot under a hedge. It is rather large, and a little like that of the wren, with a hole in front. It is made of the stems and leaves of grasses, and the lining is of leaves. Sometimes the leaves are from the beech-tree, and sometimes of ivy, according to the trees that are growing near the spot. Five or six white eggs are laid in this snug warm retreat; and when the dippers have chosen the spot for the nest, they will come back to it year after year.

The young dippers, like all other little birds, are extremely hungry, and require a great deal of attention from their parents. They stretch their heads over the nest a long way when they see the old birds coming; and if they are alarmed by anything, they will flutter out and drop into the water. Then they dive down, and come up at a safe distance.

The young dipper has a bill like his near relative the thrush, only rather more slender. But he is not destined to live the life of the thrush, so his bill alters a little as he grows older. It becomes shorter, and the tips are like those of the woodpecker's bill. His claws also get blunted with scrambling about among the stones at the bottom of the water, to look for insects and worms. Though he is accused of injuring the salmon and trout, you must not suppose he has the same diet as the kingfisher. The fact is, he does not live on fish, but on water insects and the tiny soft-bodied creatures that hide among the stones; and he has a gizzard like the thrush, which the fish and flesh eating birds have not.

THE LAUGHING THRUSH.

SOMETIMES the traveller, as he pursues his way through some of the mountainous parts of India, comes on a tract of wooded country. Here, from

THE LAUGHING THRUSH.

among the foliage, there will, now and then, burst forth a sound so peculiar that it startles and alarms him. Bursts of most unpleasant and shrieking laughter ring in his ear, and he wonders from whence they come. Perhaps, if he inquires further into the matter, he will catch sight of a large bird like that in the picture, with a white throat and head, and the rest of the plumage a reddish brown. He is called the laughing thrush, for the very reason we have stated; and the name is as suitable as any that could be given.

Numbers of laughing thrushes live in the wooded parts of this hilly district, and enjoy themselves amid the deep, cool shadow of the trees. They pick berries from the trees, and feed on insects and worms, which are procured by hunting about on the ground; and some of the species, that inhabit other parts of the tropics, are a little like the shrikes or butcher-birds in their habits.

They also live in thick woody districts, and place their nest in a bush. They are extremely noisy, like their relations, and climb about and creep among the foliage, eating the buds and leaves, and as many insects as they can pick up.

THE SHRIKE, OR BUTCHER-BIRD.

THE shrike had better have been put with the birds of prey, if we come to inquire into his private life. He has many habits that remind us of the hawk, and his beak has something about it that is very hawk-like. It is strong, and has a hooked point to it, and he uses it to tear and to devour. He will even swoop on his prey, strike it down with his bill, and carry it off, not, however, with his claws, but with his bill.

Indeed, in the old days of falconry, the shrike was looked upon as a mongrel kind of hawk, and, to quote from an old book, "thought of no great regard."

In the same old book it is said that the peasants and lower classes sometimes tame the shrike, and carry him hooded on their wrists, and let him fly at small birds.

On the continent, however, the shrike was used in catching the peregrine falcon himself. A snare was set for him, and baited with a pigeon. Then at a little distance a tame shrike was placed, fastened to a string. The shrike was to act as a sentinel, and to give warning when the falcon came near.

Meanwhile, the falconer took his ease in a hut close by, and waited the result. Presently a speck appeared in the sky, and the shrike would set up a loud scream, and run under a little shelter provided for him. The scream would be the note of warning to the falconer, and he would be on the alert, and ready to pull the string of the snare.

But, in spite of all this, the shrike has a link with the perching birds, and

THE MAGPIE SHRIKE, OR BUTCHER-BIRD.

is placed with the nightingale and the thrush. His claws are fine and sharp, and his foot is that of a perching bird. Indeed, he seems to be a link between the hawks, the crows, and the thrushes, and to partake a little of the character of each.

The shrike is only a summer visitor, when he comes to England at all, and he does not build his nest or rear his young in this country. But he carries on the trade of catching and tearing his prey, which he does in such a manner, that he is called by the rather disagreeable name of "butcher-bird."

He delights in hedgerows and clumps of trees, or thickets that are not very dense. His habit is to sit perched on a twig, or on a decaying branch, and he will remain so long that the name "excubitor," or sentinel, has been given to him on this account, as well as on account of the help he used to render to the falconer. When a small bird or insect comes near, he will pounce upon it, and kill it by blows on the head from his bill. He will then hang it up on a thorn, or on the sharp twig of a tree, that he may the better tear it to pieces; and what is left of the feast he will allow to hang up until his next meal time comes round.

In his native wilds, in Africa, quite an array of little birds and insects have been found hanging up near the shrike's nest on a row of thorns.

The nest is placed in thick bushes and high hedges, and is made of roots and moss and wool, and lined with dry grass. There are five or seven eggs, of a bluish white, spotted with brown or grey.

Though the shrike is a rare visitor to us, his race is plentiful enough in other countries. He is found in all three continents, though not in South America or in Australia. His shape is very elegant, and his costume a clear pearly grey, with the under parts white; and his wings and tail are black, tipped with white. His manner of flying is wavering, and sometimes he hovers like the hawk. The little birds know and dread him greatly.

One day a labourer, who was clipping a hedge, came to his master and said that a strange bird was sitting in it. His master went out to look, and examined every place, but could find no trace of a bird. A few days after, as he was walking by the hedge, he saw some blackbirds in a state of the utmost alarm, and uttering cries of terror. He thought a cat or a weasel must be about, but in a few minutes the strange bird his labourer had told him about flew out of the hedge and began to wheel round in the air. Sometimes it shot upwards with a kind of bound, and then hovered, suspended in the air, and moving its wings as quickly as possible.

At last it alighted on the top of a willow-tree, and then he saw it was a grey shrike.

A number of little birds were fluttering about in the utmost terror, and shrieking their notes of alarm. The shrike's attention was fixed upon them, and he seemed choosing which he should attack. Indeed, so eagerly was he watching his prey, that he did not see a gun was pointed at him. It was fired, and the little birds were at once saved from their enemy, for he fell to the ground, dead.

The shrike has several notes that resemble those of the other birds, but none of his notes approach to anything like a song. When he is alarmed, he screams loudly as the hawk does.

The bird in the picture is called the magpie shrike, and is found in Australia, where he is very abundant. He is very fierce after his prey, which he kills by impaling it.

There are a great many different members of the shrike family, but the grey shrike and the red-backed shrike are those that visit England.

THE STARLING.

THE starlings are a large and widely-spread family of birds, rather larger than the thrush and the blackbird, and yet smaller than the rook. There is but one member of the family in England, and that is our familiar friend the common starling. He is a handsome bird, his black plumage glossed with green and purple; and when he has reached his second year his beak is a brilliant yellow. He is as intelligent as he is handsome, and can be taught to whistle tunes and articulate words almost as well as the parrot.

The company of the starlings about the house and garden is always pleasant. They are very sociable, and seem to live together in great harmony. Often they feed with the rooks, and pick up worms in the field. The nest is made in the eaves of a church steeple, or in holes in walls and ruins. It is built of straw and roots, or dry grass, and the eggs are a delicate pale blue colour.

In the fenny parts of England, the young birds, when they are too large for the nests, roost among the reeds and bushes in the swampy-like districts of the country by thousands. The reeds are quite crushed down by their weight, and beaten to the surface of the water as if there had been a storm.

Along the rocky shores of the Hebrides, the starlings are found in great numbers. Early in the morning they are seen coming from their safe retreats

STORIES ABOUT BIRDS.

in the rocks and cliffs, and making their way to the fields and meadows. They are fond of the farmyards and the folds where the cattle are shut up at night, and they may be seen perched on the wall, or on the backs of the cows and horses, or else searching on the ground for insects and worms. They

THE ROSE STARLING.

keep in flocks, and fly in a compact body. When they alight on a meadow they disperse and run nimbly along, hunting for food. They dig up the worm or grub in the same manner that the crow does. All the time some of the party keep up a chattering noise, and now and then a little scream is heard. This is the note of alarm, and the starlings look up, and often fly away as fast as they can. In sunny weather they may be seen on the top of a wall, singing, in a low but sweet voice, a song that is not, of course, equal to that of the thrush, but when performed in concert is very agreeable.

The rose starling is only seen in this country at rare intervals. He is

THE SUPERB GLOSSY STARLING.

very handsome, and has a rose-coloured beak, and his plumage is glossed with violet blue. On his head he wears a flowing crest, and his back and part of the wings are a delicate rose colour. His native lands are Syria, Egypt, and Africa; and he feeds on the locust, so that in places where that scourge abounds he is held sacred. He flies in flocks like the starling, and has the same habits. And the nest is built in holes of trees and of old walls.

In India the beautiful rose-coloured birds fly in vast armies, such as we have described, and darken the air. This is just when the grain is ripe in the fields, and they play the part of the locusts themselves, and devour all before them.

THE SUPERB GLOSSY STARLING.

SOME members of the starling family go by the name of glossy starlings, and are extremely handsome. They live among woods and rocks, and have all the sharpness and sagacity of their race. They will settle on the backs of the cattle and clear them from insects. At the same time, they are fond of fruits and seeds, and sometimes, like the vulture, will feed on carrion.

The bird in the picture is the glossy starling, and is so called on account of the beautiful gloss on the feathers, that are like satin. Indeed, one member of the family has such a shine on his plumage, that the effect is like a flash of sunlight. These beautiful birds are not seen in England.

The superb glossy starling lives in the remote parts of Africa, and wears a costume of green and copper colour, the feathers being tipped with black. The throat and breast are of blue, and over all the plumage there is a gloss more brilliant than can be described. The glossy starling is a sociable bird, and flies about in company. He hides himself among the trees if he is alarmed, and makes a loud cry like a scream.

THE CHOUGH.

THE chough forms a kind of link between the crows and the starlings, and before we tell you much about him we will tell you the character that a naturalist has given to the family of crows.

He says the crows have the most varied powers, and show the greatest perfection of any of the birds. Not that they possess any one faculty in a

THE CHOUGH.

very high degree above their fellows, but they seem to have united in themselves all the qualities of the other birds.

They can soar high like the eagle, and they can walk on the ground with a firm tread. They can feed on both animal and vegetable diet, and do not refuse carrion. Indeed, their sense of smell is quite as acute as the vulture's. They have great courage, and they are very sociable, and live in communities. When properly educated they can talk like the parrot, and their cunning, mischievous ways and droll tricks show a quickness and sagacity that no other birds possess.

The chough is not very common in England. He is known from the true crows by the colour of his beak, that is red, and is different in shape from theirs. He lives entirely by the sea-coast, and walks along on the most giddy heights, thanks to his strong legs and toes.

There is a passage in the play of "King Lear" in which the chough is mentioned:—

> "How fearful
> And dizzy 'tis, to cast one's eyes so low!
> The crows and choughs, that wing the midway air,
> Show scarce so gross as beetles. Half way down
> Hangs one that gathers samphire; dreadful trade!
> The fishermen, that walk upon the beach,
> Appear like mice. The murmuring surge,
> That on the unnumbered idle pebbles chafes,
> Cannot be heard so high!"

The cliff of which the poet speaks is at Dover, and is well stocked with choughs. There is a story that they came there by accident. The whole colony are said to be descended from one pair of birds, which came from some gentleman's garden and settled there.

There are many other steep places about the coast where the choughs live; and he is so partial to Cornwall as to be called the Cornish chough, and the old Cornish families had the figure of the bird graven on their armour. He wears, as you see, the black suit of his race, glossed with blue. His beak, legs, and toes are vermilion red, while his claws are a shining black.

The voice of the chough is very shrill, and may often be heard in the places where he is found. He has even been caught and tamed, and then he was found to be a very amusing pet. He has all the curiosity of his race, and pries into everything that is at all new to him. A tame chough that was kept in a garden performed all kinds of ridiculous antics. When the gardener was nailing up the fruit-trees by the wall, the chough would peep and pry into the nail-box, and turn everything over. Then he would carry off as many nails as he could, and leave bits of cloth and shreds littered all over the walk. If the ladder were left standing, the chough would go up it with an air of intense curiosity, and hop on to the top of the wall. He would knock at the kitchen door, if he was hungry, and wanted something to be given to him. And if he could get his own way, nothing pleased him more than roaming all over the house and getting on to the roof. He disliked children very much indeed, and would hardly allow them to come into the garden. Nor would he permit the least liberty to be taken with him, even by his best friends, and he used to give them a severe peck on the slightest provocation.

THE RAVEN.

THE raven is a bird that appears to know a great deal more than he chooses to tell. Look how solemn he is, in his suit of jet black, and how intently he seems to be thinking; and he does think sometimes to some purpose. He has many good qualities, of which we shall speak by-and-by. But I am afraid at the present moment the raven in the picture has been stealing something, and is wondering where he can hide it.

A raven used to hop about the bridge over the Serpentine in Hyde Park, and was quite a public character. A lady once passed over the bridge, and chanced to drop her golden bracelet from her arm. She turned round in a great hurry to pick it up, but the raven, who was standing close by, and watching all that went on, was much too quick for her. In a second he had snapped it up, and had flown away with it in his bill, and was quite out of sight.

Of course, in process of time he came back, but without the bracelet; nor could any trace of it be found, hidden as it was in the retreat where the raven kept his treasures.

The raven's extreme gravity and sedateness impart to him a great deal of dignity. His dress of glossy black is very handsome, and has in it a shade of steel blue. He is larger than his relation the crow, and his strong beak and talons make him almost equal to a bird of prey. He is very cunning and very cautious. He is scarcely ever caught in a trap; he is far too wary for that.

But he watches with great interest while a trap is being set for somebody else, such as a fox, or a bear, or a wolf. And he does not like to lose sight of it either, for he feels a great hankering after the bait. He waits in his patient and solemn manner until some foolish creature has been caught in the trap, and then, choosing his opportunity, he will step in and devour the bait. He will also rob the nests of other birds, and carry away whatever he finds, whether eggs or chicks.

It is not easy to come near the raven when he is in his native wilds. But if his nest chanced to be found and taken, he would not try to defend it. He would stand at some distance and look on with a very mournful air, and give now and then a pitiful croak. And he has been seen to fly a long way off, and then, perhaps in an agony of grief, tumble about in the air as if he had been shot.

The raven feeds, like the crow, on the bodies of dead animals. But if he has a chance of varying his diet, he does not scruple to do so. And a taste of young lamb, or poultry, or even eggs is by no means despised. And it is on this very account that the farmer often puts a price upon his head. There is always a touch of mischief about the raven, and he likes to torment even his friends now and then.

A gentleman, who is a great friend to the birds, had a pet raven that amused him very much. He had also a pet dog, and, on the whole, the two pets were very friendly. But the raven could not refrain from playing his companion a few tricks. On a hot summer's afternoon the dog would stretch himself out in the sun for his afternoon's nap, and the raven would stand solemnly by, as if guarding his friend. But all at once, and as quick as lightning, he would give him a sharp peck. The dog would wake up with a growl, and look about him. There stood the raven, as grave and as innocent as could be, and no one could imagine he was the guilty party. The dog did not even suspect him, and after another growl he lay down again to finish his nap. But no sooner was he asleep than there came another sharp peck, that roused him up and made him very cross. This time he would look at the raven. But no, the raven has not moved a feather. There he stands, as grave as a judge, and with an air of the utmost innocence. This game would go on for a long time, until the dog lost patience, and walked away, giving up all idea of a nap.

The raven is getting scarce in England; he lives in the wild and lonely parts of the country, and is very seldom to be seen. But in the islands on the coast of Scotland, such as the Hebrides, the ravens abound; and

THE RAVEN.

here their habits and manners have often been watched by the patient naturalist.

When the raven is searching for food, he walks along the ground in the same way as the crow; but when alarmed or excited, he leaps and jumps, using the wings as well as the feet.

He flies in a steady manner, and rises very high in the air, so as sometimes to overtake the hawk, or even the eagle, as he soars along high over the mountain tops, in stormy weather, when no other bird ventures abroad.

People have even called the raven "the tempest-loving raven," but without much truth. No doubt he has scented prey at a distance, and no weather, however bad, can then keep him at home.

When he comes in sight of his prey, he alights on some stone or wall, folds up his wings, and gives an expressive croak. Then he comes a little nearer, for he is too solemn and dignified to do anything in a hurry.

He steadfastly regards his prey, looking very much as you see him in the picture. Then he leaps upon it, and begins to examine it carefully. Satisfied that all is right, he gives another croak to express his pleasure, and begins at once the feast. By this time another raven is sure to have come up, and by-and-by a gull swoops down as if asking for a share.

Before the feast is over, it often happens that the royal eagle, the king of the birds, makes his appearance, and then the revellers give way a little. They retire to a short distance—the ravens standing still and solemn, and waiting with dignified patience; the gulls walking backwards and forwards in a restless manner, and uttering peevish cries, as if angry at the interruption, yet not daring to prevent it.

If, however, the prey is large enough, a mutual feeling seems to prevail that there will be something to spare for them all, and ravens, gulls, and eagle will feed quietly side by side.

It is very curious how the ravens, as we said before, seem to spring up by magic. In parts of the world where they do not live, if an animal dies, or a fish is cast on shore, all at once, and without the least warning, there steps solemnly in the raven; then comes another, and so on, till very soon the beach or the field is black with them.

The raven builds her nest in places where no one can reach it—on some tree that juts out over the face of the rock; and sometimes the nest is close by that of the eagle, in which case the king of the birds does not seem inclined to be, on the whole, a troublesome neighbour; but when now and then little skirmishes do take place, the fault is generally on the side of the raven.

THE CARRION CROW.

OUR friend the raven, as we know, is not very particular as to his doings. He will often condescend to feed with the vultures, and has a strong touch of

THE CARRION CROW.

the vulture about him. But he is a grand and noble bird compared with his relative the carrion crow.

The very name which has been given to the crow tells you what disagreeable habits he is guilty of; and yet this part of his character is the most harmless. Nature has given him the same appetite as the vulture, and the same propensity for clearing away refuse matter. And if he kept to his trade, he would not be so cordially detested as he is. But he is a cruel and relentless bird. He watches the struggles of poor creatures who are ill or in pain, and he and his companion attack and devour them. In lonely districts he is a terrible enemy to the lambs and the sheep, and is guilty of the most barbarous deeds. In fact, he is always ready to attack any bird or animal that is exhausted and unable to defend itself.

The crow is, as you see, very much like the raven in appearance, only he is smaller. He is a solitary bird, and walks about in the same way that the raven does, and utters a harsh croak. He is also very much like the rook, though the two birds do not neighbour with each other.

The nest is on a rock or a tall tree, and is large, and made of twigs, with a lining of moss and wool and hair.

The crow chooses a crooked tree or ash that grows at the bottom of a glen or near a farmyard. There he can keep watch for any pieces of offal that may be thrown out from the kitchen door; and he is quite a weather-guide. If a storm is coming on, the crows are sure to be seen in a sheltered place, or else hurrying to some such refuge to escape what is coming.

When the crow is building her nest, she is anxious to find some soft material to line it, and she often casts her greedy eye upon the wool on the sheep's back. She is a natural enemy, as we said before, to the sheep and the lambs, and will always attack them in any moment of weakness when the shepherd is not there to protect them. But all these bad acts of hers do not prevent her from borrowing, or rather stealing, their wool. She is often seen sitting on the back of a sheep, picking out pieces with her bill, and then carrying them to the nest.

She feeds her young abundantly with all kinds of provisions, and what they cannot eat she carries away, and throws down at some distance.

A pair of crows once built their nest in a rocky glen, and a green hillock near was literally covered with egg-shells that had been thrown out as refuse.

You may suppose that the carrion crow has not many friends. There are no bounds to his rapacity. Scarcely a nest anywhere in the neighbourhood escapes him. The pheasants and partridges are dragged out of their nests, and even the young hares do not escape, to say nothing of the chickens and ducklings that he is always watching his opportunity to steal. He is just as great a thief as far as the eggs are concerned, and carries them off in his bill.

At least, he was seen to do so one day by a gentleman who was sitting reading out of doors. The crow flew over his head with the egg in his bill, and dropped it on the ground. It was not broken, as you might suppose, but was quite whole. The crow had stolen it out of the nest of the wild duck.

Like the raven and the magpie, he steals all kinds of things that can do him no kind of good, and carries them off to his hiding-place.

THE ROOK.

THE ROOK.

We need hardly describe the rook, you know him so well.

His silky, glossy plumage shines in the sun, as he struts about the field looking for worms. He is the earliest abroad of all the birds. When the dew is on the grass, and ere the sun has risen, he betakes himself to the open country, to feed on the worms that have come to the surface of the ground; or he even condescends to visit the streets and search among the heaps of refuse that lie there waiting to be carried away.

All day long you may see him at work in the pastures. He breaks up pieces of dry mould with his bill, and digs among the tufts of grass, to see if any grubs have harboured there.

Towards evening, the rooks collect into straggling flocks, and come back to their homes to roost. Every summer evening, at the same hour, they sail over the garden to their abode in the tree-tops, with the regularity of clockwork.

In hot weather they suffer greatly from the absence of moist food, and were it not for their early breakfast, would be badly off. They are seen wandering by hedges and ditches, looking for grasshoppers or what insects they can find ; and if they have young ones, their efforts to procure food are ceaseless.

In autumn all the family cares of the rook are over, and he leads a life of ease and of enjoyment. He puts on a new suit of glossy black, and wanders at large over the country. Sometimes he and his friends visit gardens and orchards, but in this case one or two rooks are always perched on the wall, to give notice of danger.

But though the staple food of the rook is worms and grubs, he has no objection to other fare. He will eat seeds and acorns and beech-nuts, and, when he can get them, eggs.

In the winter the rook is in as great distress as during a dry summer. If the snow lies a long time on the ground, nothing is to be found. The rooks then become desperate, and attack the corn-ricks in the farmyard, and do a great deal of mischief.

The distance to which the rooks fly in search of food is very great. They will go twenty miles and back by the afternoon If the rooks come back earlier than usual, there is sure to be a fall of snow or of rain the next day.

There is an old saying, "As happy as a rook on a Sunday," and some people think that rooks know when Sunday comes round. They seem to take it for granted that nobody is at home, and will venture much nearer to the house than on any other day, and take liberties not to be thought of at any other time. They seem quite at their ease, and aware that no gun is to be fired on a Sunday.

THE JACKDAW.

THE JACKDAW.

WHEREVER the rook is, there is always found the jackdaw, for they are old and familiar friends. And their habits are very much alike, for they live together in communities, and choose to reside near the dwellings of man.

But the jackdaw is more brisk and lively than the rook, and has not his solemn airs. And he makes his nest in a more sheltered position, and loves to niche it into the old church tower or the belfry. And he will place it among the giant masses of Stonehenge, in some snug corner, or even on a high cliff.

He has no idea of leaving his house open, as the rook does, to all weathers, and certainly it is much more snug. It is made of sticks for the foundation, after the fashion of the rook. And sometimes, when the jackdaw builds in a chimney, he will bring so many sticks as quite to stop it up.

You would think he would object to the smoke, but he does not seem to mind it, or care about the fire below in the grate. As for the lining of his nest, he makes it of all the soft materials he can find, or pilfer, for it does not matter to him which. Of course, wool is found there, but mixed with it are many odd things he has picked up as he goes prying about. Pieces of worsted, or bits of lace or of silk, even caps and frills, find their way to the jackdaw's nest, and make a soft bed for the little jackdaws to lie upon. And a merrier or more mischievous family you could not find anywhere.

We can give you a specimen of how the jackdaw conducts himself when he wants materials for his nest, for he is extremely sharp and clever.

There is a botanic garden at Cambridge, where all kinds of valuable plants are reared, and a wooden label is stuck in the ground near each plant. The colleges and churches in the town yield ample accommodation for numbers of jackdaws, and their nests are perched aloft in every nook and cranny. But every year new nests have, of course, to be built, and old ones repaired. Many of the jackdaws could look down into the gardens from their steeples and belfries, and they spied out the wooden labels. It seemed to occur to them that they need not be at the trouble of fetching twigs, when these little pieces of wood were close at hand. Down they came and helped themselves without any stint. Of course, the gardeners found it extremely inconvenient to have the labels pulled up, which the jackdaws persisted in doing every year, and using for the foundation of their nests. Eighteen dozen labels were taken out of one chimney only, and brought back to their owners.

We could tell you many more anecdotes of this very amusing bird.

A jackdaw once chose to make its nest on the step of a stone staircase in a church. The steps were so twisted, that the birds could not make a firm enough foundation. So they set to work and brought sticks, literally by hundreds, until they had piled them up high enough to reach a landing, and there they placed the nest.

The labour of bringing such quantities of sticks was immense.

Like the rook, the jackdaw wears a suit of black, but he is smaller than his solemn relation. His voice is shrill, and he can be taught to imitate the human voice, and talk almost like a parrot. He lives on pretty much the same food as the crows, and will eat almost anything. And, like them, he carries food home for his young ones under the tongue.

He is a most amusing pet, and his droll doings are without end. He likes to find out everything, even how to light lucifer matches. A tame jackdaw once took immense pains to perform this feat properly. When the family were gone to bed this was his favourite pastime. He had taken care to pick them up whenever he could find them, and having burnt himself severely, and lighted the kitchen fire in the middle of the night, he considered himself perfect in the art. But it was a dangerous game to play, and a stop was soon put to it.

THE MAGPIE.

THE magpie, like the raven and the jackdaw, is not very honest, and he has a great deal of cunning. He is, however, extremely handsome; his bill and his feet are black, with a shade of bluish purple. There are patches of white about him, both on the shoulder and under the body. His tail is very long, and shines with green and purple, there being a band of purple near the end of each feather, and the tip is blue and deep green.

He is a very sociable bird, and can easily be tamed. There is, however, the same objection to a tame magpie as to a tame raven—his love of stealing. Many a trinket has been snatched from a lady's toilet-table by a tame magpie and securely hidden away in his nest. What use these stolen goods can be to him, as he does not wear either rings or bracelets, is a question not easily answered. But he is not always content with this kind of booty.

On a fine day he may be seen walking in the same way as the rook.

but every now and then giving a leap in a sideway direction. His tail he carries erect, as if unwilling that its beauty should be soiled by touching the ground.

He is looking out for prey, and very soon he spies a worm half out of its hole. It is an unlucky moment for the worm, since, before it has time to draw back, the magpie has pulled it out, torn it in pieces, and devoured it. Next comes a snail, that is cleverly taken out of its shell, as a choice morsel. And, by-and-by, a still more agreeable object appears in the distance.

A hen is leading her brood of plump young chickens out into the field. The magpie spies them out with delight, and advances by a series of leaps. It is easy to guess what are his intentions, but the watchful mother is on her guard. She knows the visit of the magpie bodes no good, and, when he comes near, eyeing one of the chickens in a peculiar manner, she flies at him like a fury, her feathers ruffled, and her kind, motherly eye like a flame of fire, and scolding with all her might.

The sudden attack is too much for the magpie. He retreats before it, and flies up the nearest tree. His flight is heavy, on account of his tail, and his wings are rather short. But he is soon out of the way of the hen, who raises the shrill cry of danger, and, smoothing her feathers, begins to cluck and call her brood together.

Many poor little partridges fall victims to the magpie. Day after day he follows them about, until nearly all of them are gone. And so impudent is he that he will keep on the watch close to the farmer's house, and within sight of the family, for the chance of a hen or a duck going away for a moment, or a little chick wandering to an unsafe distance. Then he is sure to pounce upon it.

One day a farmer saw a magpie carrying off a chicken in his very sight. He fired his gun at her, but, though one leg was shot off, she still flew away and escaped. For a short time nothing more was seen of the magpie. But she appeared again on the scene in the course of a week or two, and began the same game of thieving. One day the farmer saw her going after a young duckling. It fled to the pond to escape her, and swam away on the water. The magpie seemed bent on its destruction, and rashly ventured a little way into the water. Her wings became so wet that she could not all at once rise in the air. The farmer was close behind, with a stick in his hand, and the magpie fell a victim to her habits of thieving, being killed on the spot.

THE MAGPIE.

The magpie has a chattering note peculiar to himself. He is a great talker, and sometimes turns his talents to account. If a fox or a cat is lurking about, he utters a warning cry, and keeps on doing so until the enemy has slunk out of sight.

THE MAGPIE.

He himself has many enemies, and requires all his activity and cunning to keep out of danger. The habit he has of robbing the nest of the pheasant, or the partridge, or the grouse, makes the gamekeeper at war with him, and he fires at every magpie he sees. The farmer is not any more fond of him than the gamekeeper, and his gun is often aimed at him. He remembers the magpie's love for young chickens, and that the farmyard is none the better for his visits.

Besides this, the other birds are not very friendly with the magpie, and do not much like him, because when they are away he will come and steal the eggs out of their nests.

Yet, in spite of all, the magpie is resolved, like the rook, to live near to the dwellings of man. He can pick up so many treasures, and eke out his supply of food by what he finds.

Early in March the magpies begin to build their nests, and choose the top of a tall tree, such as an ash or an elm; or, where such accommodation is not to be had, they will even place the nest in a hedge.

It is a very large nest, and can be known at once by its size and its oval shape. First there is a layer of twigs, and then a layer of mud; and then it is covered with a roof, or dome, made of twigs, and a hole is left in the side for the magpie to get in.

Within this sheltered retreat the eggs are laid. They differ much in colour, and are sometimes blue, specked or spotted, and sometimes of a pale green.

THE NUTCRACKER.

VERY rarely, indeed, is the nutcracker seen in England. Two instances only are recorded of the bird being shot as a specimen.

He is about the size of the common jay, and his plumage is a dull reddish brown, with white spots.

His home is on the continent, and among the pine forests of mountainous regions. His food is fir seeds and nuts. He fixes the nut, or cone, in a cleft of the tree, and then splits it open with a blow from his great strong bill, and gets out the kernel. He can climb and walk with the greatest ease, but does not fly with the same agility. The voice of the bird is very harsh and disagreeable, and he is far below the crows and ravens in sagacity.

The nest is placed in the deep recesses of the pine forest, almost beyond the reach of man. It is made of the young branches of the fir, and the bird seems to have a taste for ornament. She mixes in leaves from the fir to give a better effect to the nest, and the same thing is done on the inside. She

THE NUTCRACKER.

works in moss and leaves, so that the appearance is highly finished. Some pale blue eggs, spotted with brown, are laid in this elegant abode.

Besides the supply of nuts which the birds procure in the forest, they have other articles of diet. They will eat insects, and not content with this, will attack small animals, as the birds of prey do; the merry little squirrel, as he runs among the branches, often falls a victim to the nutcracker. The bird seizes it by the neck, and breaks its skull with his bill.

THE BIRD OF PARADISE.

FAR away from England are a number of islands lying in the very heart of the tropical seas. They are to the south of Malacca, and if you look on the map you can easily find them. The largest of them is called New Guinea, and is one of the biggest islands in the world.

The people who live in it are quite uncivilised, and it is not often that the white man pays them a visit. But a very beautiful bird lives in the deep recesses of the far-off forests of these islands, the most lovely, perhaps, of the whole feathered family. He is called the Bird of Paradise.

The chiefs of New Guinea, and some of the other islands, used to trade in Birds of Paradise, and sell them to a number of traders who came sometimes to buy and sell.

The traders were not white men, but came from China or Malacca, or some of the other islands, and held a kind of fair, that lasted for some time; and they always took away some choice specimens of the Birds of Paradise. Now and then one found its way to Europe.

We can give you no idea of the bird in the picture by mere description. It is called the Red Bird of Paradise, and his plumes are crimson, tipped with white. The throat is a rich green, and there is a tuft of green feathers on his head; and there are two long quills, a little like whalebone, that hang down with a curve.

His home is in a little island close by New Guinea, and every year a number of Birds of Paradise are sent as a tribute to the chiefs of another island. The native goes into the forest with his bow and arrow to shoot them. He lies hidden very snugly in a little hut he has made, and presently he sees the most wonderful sight you can imagine.

First one Bird of Paradise, and then another, comes to settle on the broad leafy top of a tree that grows close by the hut. There is the Great Bird of Paradise, the only one we ever see in England—that is, as a stuffed bird; never as a living one. You have no idea of his beauty when he is moving about near

THE RED BIRD OF PARADISE.

the tree, the sun flashing on his wonderful colours. His head and neck are yellow, and his throat of a lovely green; then his plumes are of an orange-gold colour, and look like fans of gold.

The Birds of Paradise are very lively in their dispositions, and they have quite a frolic at the top of the tree, and fly about and wave their splendid plumes until the tree seems alive with them.

But the native, from his little green hut below, has watched all their movements. His arrows have blunt points to them, for he does not wish to ruffle the plumage of the bird ; and he takes aim, and shoots at one of the merry, frolicsome group. The poor bird falls stunned to the ground, and then it is picked up and killed. When the native has killed as many as he can carry, he takes them home to his hut.

He is a native bird-stuffer, and he dries and prepares the body so that it shrinks almost to nothing. But the beautiful plumes display themselves to great advantage. Then, when all are ready, he sells the birds to the traders who stop at the island.

There are many different kinds of birds, all clad in the most brilliant costume, that live in the deep forests of New Guinea, and have never been seen alive by the white man. In that part of the world they are as common as the blackbird and the thrush are with us—and there only are they found.

THE GOLDFINCH.

THERE is a family of bright, cheerful, active little birds, that are favourites with everybody. They have short, thick, and very strong bills, that are employed without ceasing from morning till night. Both parts of the beak are thick alike, and exactly the same size, so that when the bill is closed it looks like a short cone, and it opens very wide.

Some of the family have beaks as large as their heads ; but these are tropical relations of our own finch, and are never seen in England.

We have many kinds of finches in England. They wear different costumes, and are admired for their clean, neat, bright appearance, and the pleasant song they can sing. They have a very great amount of intelligence and docility, and can be taught all kinds of amusing tricks.

The goldfinch is the prince of his tribe, and by far the most admired of any. He is a rather slender bird, with a beak of moderate thickness. He

GOLDFINCHES.

wears a little black velvet cap that comes down over his cheeks, and there is a scarlet patch over his forehead.

His back is brown, and contrasts with the white of his breast; his black wings are tipped with white, and there is a broad band of rich golden yellow across them which has a very handsome effect.

The goldfinch is not very common in England. In the early part of the summer he sometimes comes to the garden or the orchard to build his nest. It is made of fine twigs, generally of the fir, and bents of grass, pieces of wool and worsted, as well as feathers and hairs.

The birds fix their nest in the fork of a tree, such as the pear or the apple; but sometimes they choose a hedge, or a very thick bush.

In the picture the goldfinches have finished their work, and four eggs, of a pale blue, with a few pale purple and brown spots, are laid in it.

By-and-by the task of feeding the young will begin; and then the parent birds will pick off thousands of caterpillars from the fruit-trees in the garden. The whole tribe of finches are of the utmost use to the gardener, and as far as they can, help to keep down the ravages of insects. And they do more than this.

When the young finches are strong enough, they go with their parents into the fields and commons, and feed upon the seeds of many troublesome weeds, such as the dandelion and the thistle. Their fine bills enable them to pick out the seeds with ease.

In the latter end of the year the goldfinches fly about in parties of not more than twenty; but when the spring comes they separate and choose their partners.

The goldfinch will often attach himself very strongly to his owner. A lady once kept a goldfinch in a cage, and made a great pet of him. He became so fond of her that he never liked her to go out of his sight. If she left the room, he would flutter about, and try to follow her; and when she came back, he expressed by his manner the greatest delight. She often put her finger between the bars of the cage, and he would come and rub his bill against it, and make a great fuss; but if a stranger did the same, he would be highly indignant, and give the offending finger a severe peck.

The goldfinch has a great many relations, both in England and in different parts of the world.

There is one of the family, called the rice-bird, that lives in tropical countries, where people grow rice as we grow corn. When the plants are getting ripe, the rice-birds, as they are called, come and do a great deal of mischief.

It is not easy to drive them away, because the rice plants grow in mud and water, and boys cannot run about scaring the birds as they do in England. But another plan is adopted.

There are curious little huts set up in the field on long poles, and a man sits in each of them. A great many strings go from hut to hut, with bits of paper tied on them to serve as scarecrows. The men in the huts keep pulling the strings, and making the bits of paper dance up and down with a rustling noise.

Every time the string is pulled, a flock of birds rise up and fly away; but they soon come back again, and contrive to eat so much that they get very fat indeed.

This is rather fatal to them, for the native thinks they are very nice to eat, and kills and cooks as many as he can.

The rice-bird is often painted by the Chinese on his rice-paper, and a rather grotesque figure he makes of it.

THE CHAFFINCH.

THE chaffinch is much more common than the goldfinch, and is seen everywhere in the gardens and the orchards. He is almost as familiar as the sparrow, but wears a gayer costume. The upper part of his head and neck is a greyish blue, and he has a black band on his forehead. His back is of a reddish brown, and his breast has a purple tint; and, altogether, though not so brilliant as the goldfinch, he is a handsome and attractive bird.

The mother chaffinch is much smaller than her mate, and not so gaily dressed.

The chaffinch has no song of any importance, but he utters a note that sounds like "tweet, tweet," as quick as possible, and sometimes with rather a musical sound. But his melody, such as it is, soon becomes lost in the richer harmony of the other birds.

CHAFFINCHES.

When the little ones are grown up, and all family cares are over, the chaffinches fly in flocks with the sparrows and other small birds. They look for berries in the hedges and gardens, for it is now autumn, and there are neither seeds nor caterpillars. And they visit the farmyards and pick up the grains of corn, and get them out of the husk with their bills.

Often, when the farmer walks into the stackyard, thousands of sparrows and finches fly out of the stacks in clouds.

In Sweden the hen chaffinches, by going away in the winter, and leaving their mates behind, are quite by themselves. This is the reason why the great naturalist Linnæus gave the chaffinch the name of *cælebs*, a word which means a bachelor.

THE HOUSE SPARROW.

There is also another saying about him that has passed into a proverb. When a person is very smartly dressed, he is said by the French to be " as gay as a chaffinch."

THE HOUSE SPARROW.

There is a large family of birds the most familiar to us of any. They are so bold that they will hardly move out of our way, and if they do fly off a few paces, they sit down again and look as if they were not in the least afraid.

It is scarcely necessary to describe them, for everybody knows the sparrows.

Everybody knows that the sparrow is a small, stout, active, and sometimes very noisy, bird; not clad in gay plumage, like some of its neighbours, but very happy and contented in its humble station, and generally in good spirits.

As a rule, it is rather social, and likes the company of its fellows. Little parties of sparrows are often seen feeding and roosting together, but they appear to have met by accident, and any little event disperses them.

The sparrow resolves, in spite of every discouragement, to build near to the dwellings of man.

It knows where crumbs and choice morsels are to be found, thrown out into the street, and is always ready for them.

At night it roosts under the eaves, or in holes and crevices, or amongst the ivy on the wall. In summer time it does not care so much for crumbs, as plenty of food is to be found elsewhere. It visits the fields of standing corn, and later in the year haunts the stubble field.

It loves the seeds of many kinds of plants, such as the chickweed and mouse-ear; and as to the pea, it devours it greedily. On this account, the presence of the sparrows in the garden when the peas are getting ripe is not desirable.

All kinds of means are employed to drive them away, such as nets and scarecrows, and many other devices; but nothing seems to daunt the sparrows.

We have ourselves done battle with them in this respect, but they got the mastery.

As soon as the peas were ripe, the sparrows came from all quarters and set us at open defiance. They perched on the scarecrows, and with great dexterity crept under the net, taking care not to be entangled. And so intent were they on the peas, that they allowed us to approach within a yard or two, and then, with great reluctance, barely moved out of our reach; in fact, it was a game which should have the most peas, and the sparrows won it by a large majority.

Very seldom, however, does a sparrow pursue an insect on the wing, for it is rather a clumsy bird, and though it may dart again and again, generally loses it. It dearly loves to feed on house-flies, and comes seeking after them close to the house. If it were not for the sparrows and the robins, these troublesome insects would abound more than they do.

In dry weather, and when the sparrow is making its little hops and jumps on the ground, or among the branches, it may be said to look its best. Its feathers are brown and compact, and we might almost call it handsome. But when it is sitting still, and in wet weather, its plumage seems loose and untidy, and it presents a very forlorn appearance.

TREE SPARROW, AND HOUSE SPARROW.

The country sparrow, we must tell you, is much better looking than its relation in the town. The town sparrow gets so dirty with smoke and dust, that nothing can be more dismal than its appearance.

At all times the sparrow has rather untidy habits, and loves to roll in the sand or gravel on the road, and cast it up all over itself, and continues to do so for a long time; and when the sun is hot it basks in it, and sits crouching on the roof or the wall as if it were thoroughly enjoying itself.

The house sparrow cannot sing; it can only make a loud noisy chirp. And yet so great is the power of imitation, that a sparrow, that was once shut up in a cage with a linnet and a goldfinch, learnt a kind of song that was a mixture of the notes of each.

In spring, when the sparrow is choosing a partner, and thinking about building a nest, a great many fierce quarrels take place. A dozen or more sparrows may be seen scolding, and pecking, and chasing each other about, and so intent are they on their squabbles, that they allow you to come close up to them without seeing you. But when an enemy is really at hand, and can be spied by all, the noise stops in a moment, and the birds forget their quarrel, and fly off to the hedge or the trees for shelter.

The sparrow chooses a very snug place for its nest—in some chink in the wall, or under the eaves of the house, or among the thatch. When the spot has been chosen, the two birds set to work and get all the materials they can. From morning till night they carry up straws, or the withered stalks of plants, bits of rag, that may have been thrown out, pieces of thread, and feathers, or anything they can procure.

The sparrows are at all times keenly alive to their own interests, and on the watch for all they can get. They seem to know the sound of the mower's scythe, as it rustles through the long grass. The moment it ceases, we have seen them come down in flocks, and fly off again through the air, with long streaming pieces of grass in their beaks.

The sparrows are very audacious birds, and can make themselves at home anywhere. They do not always take the trouble to build a nest—that is, if they can find one empty close at hand. Taking advantage of the absence of the swallow, they get into her nest, and set up housekeeping as if it belonged to them.

By-and-by the swallows return from their winter's trip, and come back to their old haunts, intending to begin housekeeping themselves.

But the saucy sparrow puts out its head and insolently defies them. We have seen the comfortable nest of the swallow taken possession of several times in this way, until the swallows gave up the contest, and the sparrows fairly established themselves.

But they do not always escape so easily. It chanced on one occasion that the swallows came back and found their house occupied by a sparrow. The sparrow put out its head and pecked and scolded, and behaved with its usual insolence; and the swallows, as if not choosing to contest the point, went away.

But presently they came back with several other swallows, each with a pellet of earth in her bill. They began to wall up the nest bit by bit, with the sparrow inside, as if they were bent on punishing it. Indeed, they were actually seen to shut the sparrow up in its snug tomb, where it perished for want of air.

We must give some credit to the sparrow for the care it takes of its young. The nest is soft and warm, and of a good size. The mother bird lays from four to six eggs, of a greyish white colour, with spots of grey and black. When the young birds are hatched, the parents feed them all day long. Millions of caterpillars go to supply the hungry little mouths; and this is what makes the sparrow, with all its faults, a good neighbour.

The young sparrows are sometimes in too great a hurry to come out of the nest, and now and then one falls to the ground, or gets accidentally pushed out. Then it gets picked up by the great enemy of the birds, the cat, or, what is almost as bad, by children. If, however, the little sparrows leave the nest in safety, the parents do not all at once forsake them. They keep near them, and feed them for a short time. But the sparrow has more broods than one in the course of the summer, and so the young birds are soon left to shift for themselves.

In the picture you see another kind of sparrow, called the tree sparrow. It is smaller than the house sparrow, and not so common in England, though it is plentiful enough abroad.

THE BROWN LINNET.

THE linnet belongs to the family of the finches. He has the same round head and strong short bill, and the same love for picking up seeds and grain. He is no great musician, but he has a pleasant little song of his own, that he pours forth from some twig of furze on the mountain side, while his mate is brooding over her young.

His nest is in some bush of heath, and is very neat and pretty, and is made of grass and moss and wool, and lined with hair. He is very fond of the hilly parts of the country, where there are thickets of broom and heather. And he delights in the wild and craggy glens, where he can rear his brood away from the haunts of man.

He has three relations—the mountain linnet, and the red pole, and the

THE BROWN LINNET.

green linnet. He himself is the largest of them all, and the most robust. His plumage is slightly glossed, the upper part of the head streaked with greyish brown, and the back of the same colour, only of a redder shade. The feathers of the forehead are red, tipped with brown.

He is found at all times of the year, both in England and Scotland; but towards the end of autumn the linnets collect into flocks, and visit the farm-

A STORY ABOUT A LINNET'S NEST.

yards to pick grains, on which they have to live at this season. They are seen flying in little flocks, and as they fly utter soft and mellow notes. They change their plumage and spruce themselves up once a year.

THE SKY-LARK.

There is a pretty little story about a linnet's nest. It was carried home by some children, and was full of young linnets, that they hoped to rear. The parent birds fluttered round the children all the while they were carrying the nest home, and never left them. The children put the nest by a window, which they left open, that the old birds might get in. The linnets, upon this, began to fly backwards and forwards to the nest, and to feed the young. The nest was then put in a cage, but the parent birds came to it the same as usual, and perched upon it, taking no notice of the people in the room. Everything would have gone on well, but that the cage was left out of door one night, and the poor little linnets were drowned with the rain. The old birds came as usual, and lingered round the house for several days, as if wondering what had become of their little ones. At last they gave them up, and went away.

THE MOOR-LARK.

THE LARK.

WHO has not rejoiced to hear the song of the lark? On some fine clear morning, up he springs from the corn or the clover field, and as he rises, pours forth a flood of joyous music. It is a song peculiar to itself, made up of all kinds of notes, varied in quick succession. As he rises he sings all the sweeter, and up he goes higher and higher, till at last you see him a mere speck in the sky; but still ever and anon there comes to your ear a note of his wild fitful melody.

Though he rises so high, he is sure to have his nest on the lowly ground. It is hidden down deep among the springing corn, in a little hollow place scraped on purpose. It is made of stalks and blades of grass, and lined with slender fibres. The eggs are four or five in number, and are of an oval shape, and of a greenish grey colour, speckled with brown. When the mother

THE DESERT-LARK.

lark is sitting on her eggs, she would not move even if you passed close by, and may even be caught by the hand.

If she is forced to rise, she flies away close to the ground, in a tremulous kind of way, and alights as near as she can. When she brings the food to her young, she hovers over the nest, and then drops down a little distance from it.

The claws of the lark are very long; as he does not roost in the trees, but on the bare ground, and never uses them for scratching, it was a matter of inquiry among naturalists why they should have been given.

J

But a naturalist thought he had discovered the reason.

The nest of the lark is liable to be injured, placed as it is in the grass. The mower's scythe often passes over it, or even the cattle as they graze might destroy it. But the parent birds have been seen in this case to remove their eggs to another place, by means of their long claws. And the bird can also walk among the grass with much more ease.

The lark, much as we cherish him, has many enemies. His lowly dwelling is quite open to the attack of the weasel, as it prowls along the ground, and you may often have seen the hawk hovering just over it.

A lark under these circumstances utters cries of the utmost distress, and his neighbours sympathise in his grief.

The lark is a very affectionate parent. A man once took a pair with the nest of young ones, and put them into a cage. The old birds soon began to feed the brood, and continued to do so until they were able to peck. Another person had a tame lark that was very devoted to his young ones. And he not only reared his own brood, but acted as stepfather to several broods of young linnets that were put into his cage.

The song of the lark, though it seems so wild and fitful, is arranged according to a certain method by the musician.

Those who have studied the subject can tell whether the bird is rising or descending. There is first a lively air, that increases in volume as the singer ascends. Then, when he reaches his utmost height, it becomes moderate, and divides into short little passages, repeated several times over. He performs the finale as he descends, and half way down ceases to sing.

The first part of the song expresses eagerness and impatience; the second composure and calmness; the third a gradual dying away of the music.

The lark also seems to keep time by the vibration of his wings.

If there is any wind, he bounds upwards in a direct line, bound after bound. Then he poises himself in the air, his breast opposed to the tempest. But should the day be calm, he ascends in circles, and comes down in a zigzag manner.

The music of the lark stays with us longer than that of any other bird. He begins to sing quite early in the year. Our spirits are cheered, when the winter is well over, by the sound of his gladdening lay, on a sunshiny morning. And he sings to us till late in the autumn.

He ceases his song, however, earlier in the day than any other bird; and amid the summer evening choirs, the note of the lark is missing.

Sometimes the songster perches on the ground, or on a clod of earth, or even crouches among the grass, and pours forth his lay; but, as a rule, it is otherwise.

"Hark, hark, the lark at heaven's gate sings."

Towards the winter the larks assemble in flocks, and keep pretty much by themselves. When the weather is mild and open, they pick up what seeds they can find in the stubble fields; and they eat a great deal of sand and gravel to assist their digestion.

Their food at this time consists almost entirely of seeds; and they keep to the ground, and fly about in curves, gliding when on the ground in a crouching kind of way. You may walk close up to a flock of larks, before they will give themselves the trouble to rise.

THE CANARY.

IF you saw the canary bird in its native country, you would scarcely know it. Its home used to be some beautiful islands called the Canary Islands, and there it wears a costume of dusky grey.

But it is many years since canaries were brought into Europe, and their original habits and home are almost forgotten.

The birds that we see in cages come from Germany, where they are reared in great numbers. At one time a large sum of money was asked for a canary, and very few people could afford to have one. But now almost in every house you see the petted canary.

It belongs to the tribe of finches, and has a loud, rather piercing voice, and is considered to be an excellent singer. It can continue its song for some time, without pausing, as it were, to take breath, and it can open its song with the notes of other birds, such as the nightingale and the sky-lark. But in its native state its note is too harsh to be agreeable, and almost deafens the hearers.

In Germany a great room like a barn is prepared for the canaries, and at the end there are several openings, where trees are planted; seeds, such as groundsel and chickweed, are thrown upon the floor in plenty; and every possible material is supplied that the birds will want to make their nests.

THE WILD CANARY.

In England the tame canaries often make their nest in the cage, and rear their little brood, to the great delight of their owners.

A lady had some tame canaries that she hoped would rear their young ones. She thought she would make the nest for them, and save them the trouble. But the bird was not at all satisfied with her workmanship, and tore the nest to pieces with her beak, and scattered it about. Then she set to work, and made it over again.

The tame canary is always a captive, but a happy one. It has never

THE TAME CANARY.

known liberty, and does not therefore regret the loss of it. It is very affectionate, and becomes so familiar that it will hop about the table, and perch on its mistress's shoulder.

It lives a long time. One that was hatched in a cage, lived with the same mistress twenty-six years.

THE CUCKOO.

THE order of climbers includes some birds that can hardly be said to climb, though they live much among the branches of trees.

THE CUCKOO.

There is a large grey bird that comes to us in spring, just when the leaves are budding, and the meadows are beginning to be covered with flowers. When we hear his well-known note we know that summer is near. His song has no great variety in it. It is but a repetition of the well-known word, "cuckoo! cuckoo!" but perhaps no other note gives us more pleasure,

THE CUCKOO.

He is the lowest of his tribe as regards his climbing powers. Nor does he use his bill in the least, like the parrot, for helping himself to climb or for clinging; nor, like the woodpecker, for digging into the bark of trees. Yet, after his fashion, he can climb about on the trees, though he cannot mount up

THE JAY CUCKOO.

the stem. The cuckoo is, in fact, half a climber and half a percher, and it seems rather foolish to have put him among the climbers at all.

His bill is of a moderate length, and a little curved at the end. His wings are short and his flight is feeble. Nor does he, in his native woods, take any long journeys, except from one tree to the other. He alights on the highest boughs, and begins to hunt about among the foliage for insects, threading the most tangled mazes, and hopping from one bough to another without opening his wings.

All the insects and caterpillars that lie in his route are, of course, snapped up and devoured. His movements are very quick; and his tail is long and helps him to balance himself on the boughs—indeed, the Indian gives him the not very elegant name of "cat's-tail."

He is a tropical bird, and lives both in Africa and America. His costume is rather sober, but with a pleasant lustre; and his long tail is often barred with black and white. Nay, in some instances he wears a splendid plumage of emerald green.

His food is insects and caterpillars, and grain, with berries and fruit; and he will even attack mice and lizards, and eat the eggs of other birds. At certain times of the year his short wings have to bear him a long way, for every summer he pays us a visit; and we expect him as confidently as we look for blossoms on the fruit-trees or the hawthorn on the hedges.

"In April come he will.
In May he sings all day.
In June he changes his tune.
In July he may fly.
In August go he must."

The mother cuckoo, perhaps on account of her wandering life, or for some other reason which we cannot find out, never seems inclined to build a nest, or, indeed, to trouble herself with family cares. Yet, like the other birds, she has a family to provide for—or rather, eggs to hatch. At this time of the year her feathered neighbours have finished their nests and laid their eggs. And the cuckoo seems well acquainted with the fact. As she goes her rounds among them some fine morning, she looks sharply about her, and makes up her mind what to do.

There are some of the birds that she chooses as foster-mothers for her own offspring. She does not consult them about it, for perhaps they would refuse—for it is no great honour—and mischief, as a rule, comes of it, that is, to their own poor little nestlings.

But the birds are in happy ignorance of the honour intended for them. The cuckoo flies stealthily about among the bushes, and visits first one nest and then the other while the parents are away. She is very cunning, and seems to know exactly the home that will suit her offspring. It must be fed on

THE GOLDEN CUCKOO.

caterpillars and grubs, and she chooses the little birds that provide such fare for their families. She fixes on the hedge-sparrow, or the lark, or the blackbird, and contrives to drop an egg in the nest. Her egg is small and goes in very easily. But she has often been accused of destroying the other eggs in the nest, and we are afraid with truth. When she has accomplished this feat, she flies away, and drops another egg in another nest, and so on.

By-and-by the parent birds come back and find the strange egg in the nest. Sometimes they are very angry, and turn it out. But more frequently the mother bird takes it under her care, and sits upon it with her own.

After a proper time the eggs are all hatched, and the strange little bird comes out of its shell, and the mother begins her usual work of feeding. But it is a curious fact that very soon her own offspring disappear, and the young cuckoo remains master of the nest. It is, in fact, so large as to take up all the room.

In the meantime, the foster parents do not seem angry with the behaviour of the cuckoo, but feed it constantly. It requires more food than all their young ones would have done put together. But they labour all day long on its behalf.

When it is able to leave the nest, they keep near it, and protect it from the other birds, which seem to know that it is a cuckoo, and show their dislike by teasing it.

The cuckoo flies in a swift gliding manner, at no great height. Sometimes it skims over the ground, and, alighting on some stone or crag, balances itself with its tail, and begins to utter its note. It can walk on the ground after a fashion, but with no great ease. When on the trees it clings to the branches and climbs among them, searching for insects. It can limp round a bush and peck the worms and grubs from it, and destroy myriads in a very short time. The young cuckoos remain until September.

The kindness displayed to the young cuckoo by the mother birds is very curious. One day a young cuckoo was put, by a naturalist, who wanted to watch what would happen, into the nest of a tit-lark. There were three young larks in the nest, and when he came back to look the next day, he expected to see the young cuckoo turned out. But a curious scene presented itself to him. The poor little tit-larks had been thrown out, and lay dead close by the nest. Within the nest was the young cuckoo, sole possessor, and the mother tit-lark was hovering over it with outstretched wings, to shelter it from the rain, which was falling in torrents.

We must, however, not forget to say that when the young cuckoo is about fourteen days old its desire to turn out the other little birds goes off.

THE PARROT.

THE climbing birds do not catch their food quite in the same way as the perchers. In fact, their food is in rather a different place.

Sometimes it grows upon a tree in the shape of a nut or a fruit, and the climbing bird makes his way among the branches as nimbly as the monkey does. His feet are made for climbing, and his toes are arranged in pairs, two in front, and two behind, and the outer toe turns backwards like a thumb. Indeed, the foot can hold a fruit or a nut as well as if it were a hand.

We are speaking now of the parrots, that form a class by themselves. They are such famous climbers that they have often been compared to the monkeys. And the parrot, like the monkey, rarely walks on the ground, but climbs, and makes his way from bough to bough, keeping always in his leafy home in the tops of the forest trees.

His expertness in climbing is so wonderful that we must pause a few minutes to watch him.

He has a hooked bill, which is very strong, and he uses it, as well as his feet, to climb with. He lays hold of the branch that is over his head and hooks himself on to it. And then he raises up his body and grasps a branch on either side, and thus makes his way along.

His bill is of the utmost use to him; without it he could not crack the forest nuts and get at the kernels. And he is extremely fond of all such diet. He can move the upper part of his bill as well as the lower, which gives him a great deal of power. We must tell you that all birds can do this in some degree, but the parrot better than any. The upper part of his bill is not a mere piece of the skull, but is quite separate from it, and connected by a joint, so that it can move with the utmost ease. And the muscles that move the bill are very strong indeed, so that it can crack the hardest fruit, and do almost anything.

The parrot has a thick fleshy tongue, that can taste and relish the food. And in some kinds of honey-loving parrots, that feed on the nectar of flowers, the tongue has a kind of brush at the end of it, made of tiny filaments, that can spread out when wanted, and sweep off the honeyed juices of the flower.

The parrots, as a family, are dressed in the richest attire. They vie with the other birds of the forest, in their costume of red, and yellow, and blue, and green. They are very sociable birds, and live in flocks, and roost in the great hollow trees of the forest. While in their native state they have not

THE GREY PARROT.

learned to speak our language, like their brethren in cages, but they can talk fast enough in their own. The noise and screaming they make is beyond description.

THE PARROT.

THE AMAZON PARROT.

    The parrots lead the most joyous life in the forest. They are very friendly and affectionate with each other, and keep all together, like a happy family.
    In the morning, when the air is cool and fresh, and the dew sparkles on

the flowers and grass, and the burning sun of the tropics has only just gilded the tops of the forest trees, the parrots are all alive. They climb about

THE WAVED PARROT.

among the branches, hooking, and swinging, and putting themselves into all kinds of positions. They are getting their breakfast of fruit and nuts and berries, and a pretty sight it must be to see them, their gay colours glistening among the green leaves.

THE COLLARED PARROT.

They are very fond of taking a bath in some clear stream, and they generally go at one particular hour, and are as punctual as clock-work. The bathing scene is like a frolic, for they splash about and roll over each other, and chatter and scream, and seem to be thoroughly enjoying themselves. When they have bathed enough, they fly up to the branches of the trees, and dress their feathers, and make themselves very clean and tidy; and in the middle of the day, when the heat is the greatest, they take a nap, and all is silent.

Their roosting-place is in some hollow tree, often in the hole made by the woodpecker.

The parrots get into the hole until it will contain no more, and the rest sleep close by, hooked on to the tree by their bill and claws.

The mother parrot does not take the trouble to make a nest. She lays her eggs in a hole, and all the mother parrots lay them in the same place.

Because the parrot can be taught to speak, and has such droll ways, he is much sought after as a pet.

The Indian makes a trade of catching parrots, and he goes into the forest on purpose. He uses a blunt arrow, for he does not want to kill or even to hurt the parrot; he merely wishes to stun him and make him fall to the ground, so that he can carry him away. And sometimes he will not use arrows at all, but will light a fire under the tree, and burn a kind of plant that makes a strong stupefying smell. The parrots begin to feel stupid and sleepy, and drop to the ground as if they were dead.

Sometimes a curious and rather cruel thing is done to the parrot.

By nature he wears a green dress that you would think was handsome enough; but the Indian fancies he can improve on nature. He tries to get the parrot when he is young, and his green feathers are only just beginning to grow. He takes off the feathers, and rubs the skin with a kind of dye that changes the colour, and when the feathers grow again they are not green, but red and yellow.

The parrot in his new costume is thought a great rarity, and is worth much more than if he had been let alone. But he is never so cheerful and lively as he used to be, and has rather a drooping and dejected air. In fact, his health is injured by the process.

THE PARRAKEET.

THE parrots have a great many relations in the tropical parts of the world. There are parrots everywhere in the woods and forests of hot countries, and their shining colours are in harmony with those of the humming-bird and the sun-bird.

THE GROUND PARRAKEET.

The parrakeet is, perhaps, the most beautiful of the whole tribe.

The emerald green of the body, and the deep red of the beak, and the rose-coloured collar round the neck, form a picture of beauty that must be seen to be fully appreciated; and the two long feathers that you see in the tail are an exquisite blue.

K

Nature in those sunny lands seems to delight in the most brilliant hues and tints.

THE GAKUBA PARRAKEET.

There is a little difference between the parrot and the parrakeet. The lower part of the bill is, as you see, short and notched, and the claws are more slender and not so strong as those of the parrot.

THE PARRAKEET.

There is a beautiful parrakeet, called the ringed parrakeet, that is often seen in a cage in England, and is a pet bird greatly admired. It has very slender feet, and can run along the ground—a habit peculiar to the parrakeets, and one of the reasons why they differ from the parrots.

In the Spice Islands there is a small green parrakeet that is much sought

THE DAPPLED LORIKEET.

after by the native. He does not care for it as a pet, but as an article of food. The parrakeets have most dainty fare, for they live on the berry-like fruit of the spices, and their flesh has a very delicate flavour. When they are fat and in their best condition, the hunter takes a walk into the grove on purpose to shoot them.

The green bower in which the birds are sitting is so like the colour of their wings that they are not easily seen. Indeed, if they kept quiet, nothing would be the matter. But it is not in the nature of a bird to sit still many minutes. They soon begin to move from bough to bough. The fact is, they have eaten all the berries from one bough, and want to go to another. Then the hunter can hear the rustling of their wings, and can see them quite plainly. Of course, he takes aim with his gun, and shoots as many as he likes.

We told you that one kind of parrot has a brush-like tongue, and that he is very fond of honey. He is called the lorikeet, and is very splendid with his red and yellow body, and blue head, and yellow tail.

He climbs and hooks just as the parrot does, but he cares little for fruit; all he wants is the sweet sugary juice of the flower, and he sweeps it off with his tongue. No other member of the family possesses such a tongue, for it would be of no use to them.

The lorikeet lives in Australia, and spends all his time amid the beautiful flowers and blossoms that abound on every hand. He is so intent on his banquet that he cannot see what is going on quite close to him; and the hunter with his gun often comes out to look for him. The flesh of the honey-loving bird is a great dainty, and the poor lorikeet often falls a victim. Indeed, being caught in the very fact, his crop is full of nectar, and the native takes up the dead bird and sucks the rich store out through the beak.

THE COCKATOO.

THERE is a very interesting branch of the parrot family that wear a crest of beautiful feathers on their heads. They can set the crest up and down just as they like.

They are called cockatoos, and, like the parrots, they live in the forests; and, like the parrots, they are not always out of mischief; for we did not dwell on this part of the parrot's character at the time we were speaking about him.

The parrot is not always satisfied with the feast provided in his native woods; he often pays a visit to the orchards and the stackyards. There is a North American parrot that sets no bounds to his thefts and pillages. He

THE NORTH AMERICAN PARROT.

THE BLACK, OR RAVEN COCKATOO.

and his companions come and fall upon the fruit-trees like a sheet of colour; for they are very beautiful in their gaudy plumage; but they eat all before them, and nearly strip the orchard bare.

In countries where the rice is grown, the cockatoos play just the same game. They come in large companies, for they are as friendly and sociable

THE NESTOR COCKATOO.

with each other as the rest of their tribe; and they settle in the rice-field, and eat, and tear, and break, till there is no end to the damage.

THE HELMET COCKATOO.

Of course, the owner of the rice-field does all he can to destroy them, and looks upon them as the greatest pests in the world.

There are many cockatoos in Australia, not black, like one about which we shall tell you presently, but of a light rosy tint, and with a sulphur-coloured crest. They like to fly about on the banks of the rivers where there are great trees close by; and here they enjoy themselves as the parrots do, and bathe, and take their afternoon naps, and lead pretty much the same sort of life.

The native thinks nothing is better sport than to go out to shoot cockatoos. He has to be very careful, for the birds are extremely shy, and if they catch sight of him will fly away to a distance. So when he has seen the flock of birds in the air, which he is almost sure to do if he goes in the right direction, he hides himself behind a bush. Then he creeps in the most cautious manner, and contrives to come as near to them as he can.

The birds, meantime, are going to roost on the trees, and make as much noise and uproar as the parrots. They spy out the native before long, for there are several cockatoos on the watch ready to give the alarm, and they huddle together as close as they can, and begin to be frightened.

The native has a spear in his hand that he manages in a very clever way. He flings it among the birds so that it spins about and knocks one or two of them down. Of course, they rise up, and try to fly away; but, whichever way they turn, some unlucky bird is sure to meet with a blow from the spear.

The cockatoo is, as we told you, very affectionate, and the native is unkind enough to trifle with his feelings. He picks up a poor wounded bird, and fastens it to a tree. The bird makes a piteous outcry, and its friends and companions come back to see what is the matter. Then the native throws his spear, and knocks some more of them down.

Almost all the cockatoo family are dressed in light rosy-coloured plumage. But there is a great black cockatoo that, as his name tells you, wears nothing but sable. He is a very curious bird, and lives in Australia, and also in the islands near New Guinea, where the Birds of Paradise have their home.

He has rather a small body, but his head is very large indeed, and he has a crest of black feathers. His cheeks are red, and as for his bill, you see what it is like by looking in the picture.

That bill of his can crack the hardest nut in the forest. The nuts grow on a very tall tree, and the shell is as hard as iron. No other bird can manage to get out the kernel except the black cockatoo. He holds the nut in his foot, and wraps it in a leaf to keep it from slipping, for it is very

THE BLACK COCKATOO.

THE CASMALOS COCKATOO.

smooth indeed; and then he digs the sharp end of his bill into it. He contrives to get out the kernel with his tongue bit by bit.

His tongue is the most curious part of him. It is red, and can be pushed out to a great distance. There is a horny kind of plate at the end of it.

He is not so noisy as the rest of his tribe, and when he flies he makes no sound. His note is a low plaintive whistle, and the slightest wound kills him. He does not live in company like his relations the parrots; only two or three black cockatoos are seen together, and they are not at all common.

THE BRUSH TURKEY.

THE BRUSH TURKEY.

THERE is a large family of birds, called scratchers, some of which are of the utmost use to mankind. Their flesh is wholesome, and forms an agreeable article of diet, and their eggs are a staple article of food.

This very useful class of birds includes, as you will at once perceive, the fowls that peck about the farmyard, and that are sold in the market daily; and it comprises turkeys, pheasants, grouse, and many more.

All these birds feed on grain and seeds, and are provided with a strong gizzard to crush the food.

The ground is the place where they seek their food, and their strong toes are usually occupied in scratching for it. They are careless nest-makers, and have none of the skill of the smaller birds. Often the nest is a mere hole scratched in the ground; nor can they utter sweet and mellow notes like the smaller birds. They can cluck and crow, and scream and gobble, but there is not a single musical voice among them. Some of the tribe—we might say nearly all of them—are splendidly dressed, and belong to warmer countries than our own. The turkey, though he is reared with great care in England, is not a native of our climate. He and his companions in their native state live in the forests of North America; and his plumage is far more handsome than that of his relative in England.

The brush turkeys are natives of Australia. They are very curious birds in their habits. In the spring, when the eggs are about to be laid, they set to work and get together a heap of stalks, and grass, and rubbish. It takes them some weeks to make the heap large enough. When the mound is finished, the mother turkeys lay their eggs inside it, and leave them there to be hatched by the heat of the decaying matter. But though the turkey does not sit upon her eggs, she keeps to the spot, to watch over them; and she will even put in her head to look if all is going on right.

The little ones come out of the eggs with their feathers on, and their wings strong enough to enable them to fly off to the branches of the trees.

THE CAPERCAILZIE.

THE grouse family always seem to remind us of the moors and heaths of Scotland, or the mountainous parts of Europe. And such are the places in which they make their home.

They are not so gaily attired as the pheasants, and have none of the handsome crests or the brilliant colours so much admired among those rare and beautiful birds. There is no naked skin about the head, except one small space that surrounds the eye, and which is of a scarlet hue. The tail, with some few exceptions, is short, and the hind toe small and weak.

Some members of the family are found in warm countries, but, as a rule, the grouse inhabit cold and alpine regions. And in some cases the bird is protected by a clothing of feathers down to the tips of the toes.

The bird in the picture is the prince of his tribe, and his name is derived from a Gaelic word meaning "horse of the wood," and refers to his large size. He is as large as a turkey, and his plumage is black freckled with white, so as to look almost like grey. The breast is a handsome green, and the wings chestnut red, while the tail is black tipped with white.

The northern parts of Scotland once abounded with this magnificent bird. His chief haunt was the pine forest, and his favourite food the tender leaves and shoots of the Scotch fir. But his great beauty and savoury flesh caused him to become an object of attraction. He was easily found on account of his size and appearance, and at last was completely hunted down and destroyed. The last specimen was killed about seventy years ago in the woods of Scotland.

But great efforts have been made to repair the loss. Some of the Scottish nobility have procured birds from Sweden, and set them on eggs. Many young birds thus hatched have been turned out into the forests, and in time the race may again become plentiful. Besides the shoots of the Scotch fir, the capercailzie, or "cock of the woods," as he is called, will eat juniper-berries, cran-berries, and any other kind of forest fruit, and the young birds will even eat insects and worms.

In the early spring, when the snow is yet on the ground, the bird places himself on a pine-tree and begins what is termed his love song. He not only sings, but dances. His wings drop, his tail is spread out like a fan, and he looks very much like a turkey-cock when he is angry and gobbling. The note he utters sounds like the word "pillar, pillar," and he goes on getting more and more excited until he hardly knows how to contain himself. He has his dancing times—from the first dawn of day until the sun rises, and then from sunset until dark. And he has his dancing places, where the game goes on spring after spring. And what is rather funny, the old birds will not allow the young ones to play and sing in their places. But if the old bird happens to be shot, a young one is sure to set up his note in the very same spot in the course of a few days.

The mother birds, who are scattered in the forest, hear all this singing and dancing, and they reply to it by a note very much like the croak of a

THE CAPERCAILZIE.

raven, and utter the sound "gock, gock, gock." And they come to the tree where the performance is going on.

The nest of the capercailzie is made upon the ground, and there are about twelve eggs in it, of a pale yellow brown spotted with orange. The mother bird sits a whole month, and is quite forsaken by her partner, who skulks about in the wood, and renews his plumage, while she is busy with her domestic cares.

The capercailzie lives a great deal upon the ground, unless the snow happens to be deep. He sits also on the uppermost boughs of his favourite pine-trees, and in the night roosts there. Sometimes, however, if the weather is very severe, he gets quite into the snow, and buries himself. He can fly to rather a great height, considering the tribe to which he belongs, and can keep on the wing for several miles.

The birds are sometimes kept in Sweden in aviaries, and made so tame as to eat out of their owners' hand. Their food consists of oats and the usual leaves and shoots of the pine. And large branches are put into their cages once a week. And in England the same experiment has been made, and a brace of the birds have not only lived in confinement, but reared a family of six young ones. When the young birds are reared in this manner, they become as tame as the cocks and hens, and may be allowed to go at large. At the same time, the cock never loses his native courage, and will often fly at people and peck them.

There was an old cock that for many years was well known on a certain estate in Sweden. When he heard the sound of people's footsteps, he would come out of his lair and peck their legs and feet, flapping his wings all the time. And another bird lived in a wood through which a public road passed. Whenever a person went by on the road, out would come the capercailzie, and make an attack, and it was not easy to drive him away. At length he was caught and carried from the place, but he was so fierce that the man who took him was obliged to let him go, and in a day or two he was seen at his old haunts.

In the northern forests the sport of hunting the capercailzie is thought to be excellent. It is not very easy to take aim at the bird, because he has a way of dipping down from the branch to the ground, and running away before the hunter can come within firing distance. In the winter, however, the birds are in companies of sometimes a hundred and fifty, and they keep to the sides of lakes and rivers. And then the sportsmen go out with their guns. And in one part of Sweden a very destructive plan is adopted. The birds are shot in the night by torchlight. The plan is to watch their flight as they go into the forest to roost, and to mark the direction so as to be able to follow it. Then at nightfall two men set out, one with a gun, and the other with a long pole, to each end of which a torch is fastened. The man with the torch goes first, and they make their way as well as they can to the tree on which they believe the birds to be roosting. Very soon they reach the spot, and there are the poor birds wrapped in slumber; even when they wake they seem to have no idea of escaping, but stare in a stupid manner at the blazing torches till they are nearly all shot down.

THE BLACK-COCK.

THE handsome bird of which we have been speaking is not fitted to live in the present state of our island. Britain is too thickly peopled for it to find any secure retreat, and the size and beauty of the capercailzie, and the great demand for it as an article of food, have occasioned its destruction. Though efforts are being made to bring it back, yet it belongs, as it were, to the past.

But there are still some very handsome members of the grouse family remaining, and among these is the black grouse, or black-cock, as he is called. He is a large strong bird, with handsome plumage, and the plumes of the feathers are very full and large, especially those at the hinder part of the body, which project beyond the tail feathers in the manner you see in the picture. The feathers of the head and neck and hind part of the back are glossy and smooth like silk. On the whole, he is one of the finest native birds we possess. His favourite home is the moors and hills of Scotland, and his habits are well known to the sportsman, as well as to the naturalist.

Early in the morning of a clear autumnal day he may be seen by the watchful observer threading his way through some romantic glen where the heather grows. He pecks off tender morsels of the young twigs with his bill, and as he goes on, meets with berries and wild fruit, none of which he despises. His crop is very large, and by degrees it becomes quite full of twigs and berries that have passed into it, and go down into the gizzard like a compact mass. The mill is, however, strong enough to grind it into pulp, aided by morsels of sand and gravel, which these kind of birds are in the habit of swallowing. So, in fact, the black grouse, like many of the family, goes through a process that very much resembles chewing the cud.

He does not wander far from his native haunts, and the least noise alarms him. If a footstep approaches he flies off to some secret spot, and lies hidden

THE BLACK-COCK.

until the danger is past. He is found in many parts of England, as well as of Scotland. Like his splendid relative the capercailzie, he has been driven from one spot to another by the progress of men and cities and cultivation. But there are some few spots remaining in which he dwells. Among the many plantations in Northumberland, in the New Forest, Hampshire, and on the heathery hills in Somersetshire, and the romantic glens of North Wales, he is still to be seen. But nowhere is he so abundant as in the north of Scotland, where grouse shooting has become a yearly custom with the sportsman.

The nest of the hen bird is in the shelter of some low bush or among the grass. It is made of withered grass, and sometimes of twigs; and the eggs are of an oval shape, spotted and dotted with brownish red. The bird places her nest in such a low situation, that in very wet seasons it is apt to be filled with water. The hen performs the duty of rearing the young without any assistance from her mate.

THE LYRE-BIRD.

IN the beginning of the present century, a party of rather turbulent Irishmen were sent on a voyage of discovery to the interior of New South Wales. The governor hardly knew what to do with them, and he thought the hardships of travelling in an unknown country would cure their restlessness. When they returned, they brought with them a bird which they called a pheasant. Its size was that of a common hen, of a reddish black colour, and with strong black legs. It had a crest on its head, but its tail was the most extraordinary part of it. It spread out in the shape of a lyre, and was composed of several feathers of a light brown colour, inclining to orange, and shading into silver. The end of each feather was black. The feathers were, as you see in the picture, of a different texture, alternately thin and thick.

In the mother bird the tail has not the lyre shape, and is more like that of the peacock. Nor has she the crest of her mate. Nothing can exceed the beauty of this extraordinary bird, and it is of all others the most shy.

The tail has not the dazzling splendour of the peacock, but it surpasses it in beauty of shape. There are, as you see, two large curved feathers, of black and brown striped, that curve into the form of a lyre, and between them are a number of finer and gauze-like feathers that fill up the space and give them a most elegant appearance. Nothing so striking or graceful had been ever imagined, and yet it had been hidden in the wild bushes of Australia from time immemorial.

Of all the birds the lyre-bird is the most difficult to catch sight of, much less to procure. Its large strong feet are made for running, and it is constantly going up and down among the brushwood, from the top of the mountains to the steep and stony gullies below.

It carries its tail erect, so that it can come to no danger. It has a loud cry, which may be heard a long way off, and another note, which may be called a song, but which cannot be heard unless you are close by.

The naturalist goes through unheard-of toils to catch sight of the birds. He lies hidden among the brushwood, and hears their loud shrill notes, for days together, without being able to obtain a glimpse of them. Quite determined to do so, he does not give up his point, but climbs along the gullies and ravines, where he has to cling to trees and creeping plants to keep himself from falling.

These are the spots where the birds often resort; but if so much as a branch cracks, or a stone rolls over, they take the alarm and are gone. Even when the hunter has come up with one of them, he has to crawl among the branches of the trees and remain perfectly motionless. If the bird is not singing, or engaged in scratching for food, it is almost sure to perceive him if he stirs either hand or foot, and it vanishes as if by magic.

It runs with the utmost rapidity, aided by its wings, over rocks or logs of wood, or whatever comes in its way.

It does not often fly into a tree, except to roost. It scratches about the ground and the roots of trees to pick up seeds and insects. Its nest is very large, and a little like that of the magpie. There are twelve or sixteen eggs in the nest, of a white colour, with a few light blue spots. The young birds scamper about with the utmost rapidity, and hide themselves amongst the rocks and bushes. In some places, where roads have been cut through the bush, the bird is more frequently seen, and a man on horseback can approach it more easily than when on foot. It seems less afraid of the horse than of the man. Sometimes it is pursued by dogs, that are taught to rush suddenly upon it when it leaps down from its roosting-place in the tree. And sometimes the hunter wears one of the beautiful lyre-like tails in his hat, and keeps it moving about while he hides in the bushes. The bird is taken by surprise at what he supposes to be one of his own species, and comes within reach of the gun.

Another way is to whistle, or make some unusual sound, upon which the bird will come forth out of curiosity, and allow himself to be seen; but unless

THE LYRE-BIRD.

the gun is fired in a moment, he is half way down the valley. Indeed, shooting the lyre-bird is totally different to any other kind of sport, and the most clever sportsman could do nothing unless he understood the nature of the country and the habits of the bird. The native is by far the most expert hunter of any. He likes to deck his hair with the plumage of the lyre-bird, and to glide noiselessly among the bushes with a gun in his hand. So cautious is he, and so silent, that he can always approach nearer to it than any one else, and rarely suffers it to escape.

Besides its running powers, the bird can take very wonderful leaps. At one leap it can rise ten feet from the ground. Its habits are solitary; but two lyre-birds have been seen at play, chasing each other round and round, and carrying their elegant tails in an upright position. It has also the habit of making a round hillock, on which it comes every day and erects its tail, and tramples the ground, and utters all its notes—sometimes mocking those of other birds, and even making a howling noise like that of the dingo, or native dog.

Besides its loud full call, which may be heard echoing to a great distance, it can sing the little song we have mentioned. The strain is often broken off suddenly, and then resumed again.

The nests are sometimes placed on the ledge of some projecting rock, or on the stump of a tree, but always near the ground. One of the nests which was seen by a naturalist was deep, and shaped like a bason, and lined with the bark of trees and fibrous roots.

THE PHEASANT.

THE pheasant family contains large and handsome birds, with beautiful plumage, and white tender flesh, that is much sought after as a delicacy. The male birds wear by far the most gorgeous array, and sometimes shine in plumage of gold and silver. But it is a curious fact that the hen bird, when she is getting old, will often assume the beautiful colours and gay plumage of her mate, and become a sort of natural curiosity.

Next to the peacock, the pheasant carries away the palm in beauty,

THE SILVER PHEASANT.

both for the lovely colour of his plumes, and the happy manner in which they are blended.

There is an old story told about a famous king of Lydia, named Crœsus, who was said to be the richest monarch in the world.

He was one day seated on his throne, in his royal robes, and in all his magnificence, and asked Solon, the Greek philosopher, if he had ever seen anything so fine.

It was rather a foolish question. And Solon replied, that having seen the beautiful plumage of the pheasant, he could not be surprised by any other grandeur that might be displayed before him.

Indeed, the attire of the pheasant is rich and rare. The eyes are surrounded with scarlet, sprinkled with tiny black dots. On the front of the head there are dark coloured feathers, mixed with a shining purple, and the top of the head and the upper parts of the neck are tinged with a dark green

THE BLACK PHEASANT.

that shines like silk. Sometimes the top of the head is of a shining blue, and looks blue or green according to the light in which it is seen. The feathers of the breast and shoulders, and the sides, have a tinge of exquisite purple, with a streak of gold. The tail is long, and, in the silver pheasant, is of a silver white.

The pheasant, thus grandly attired, is no less admired when served up at the table.

His flesh is so delicate that its delicacy once became a proverb, and when a doctor in those days wished to recommend an article of diet, he used to say it was as nice and as wholesome as the flesh of the pheasant.

THE COMMON PHEASANT.

There are many varieties of the pheasant, such as the spotted pheasants of China, and the gold and silver pheasants, also brought from that country.

The spotted pheasant is related to the gold and silver species. It is the most magnificent of the whole tribe, and lives in the dense forests of Java and Sumatra. Its wings consist of very large feathers, nearly three feet long, the outer part of which is adorned with rows of great spots like eyes. It is called Argus, after the imaginary Argus with his hundred eyes.

184   *STORIES ABOUT BIRDS.*

The daily life of the pheasant is very much like that of the grouse. He loves the thick plantation or the tangled wood, and during the summer and autumn has the habit of sleeping on the ground, though in the winter a tree is chosen on which to roost.

CHINESE PHEASANTS.

Early in the morning he visits the open fields, and searches for the tender shoots of the grass and of many of the meadow plants, and will pick up worms and insects. Later in the season, acorns, and beech nuts, and wild berries form articles of diet. But during a severe winter the birds require to be fed, or they would suffer from hunger. Then they become very tame, and come when they are called.

The pheasant walks about like our familiar cock and hen, and has very much the same habits. Though he can ascend a tree, yet, like them, he lives upon the ground, and runs very fast. When he is alarmed he runs off to the nearest place of shelter, or even uses his wings. In this case he flies heavily, like our common fowl.

The mother bird takes very little trouble about her nest, and merely has a slight hollow in the ground, on which she places a few leaves. Very often she declines to take any trouble about her eggs, and then a hen has to be fetched, and acts the part of a foster-mother.

THE OSTRICH.

IN the barren wastes of Africa, and also of Asia, the traveller, as he journeys wearily onward, meeting with but stunted herbage and no water, sees from afar something that alarms him. It looks like a body of horsemen scouring the desert, and, as he fears, bent on plunder. There is no way of escape, and as he looks hither and thither the dreaded object approaches. Then his heart beats more freely, and his spirits revive. The band of horsemen, as he supposed it to be, turns out to be birds. And he is not the first traveller by any means who has made the mistake, and imagined the ostrich to be a man on horseback.

In the first place, the ostrich is quite as tall, and as he runs swiftly along there is nothing at a distance that he more resembles. He always feeds in a flock, and the barren wastes have been his home from time immemorial. He eats grass, and grain when he can get it, and does not seem to care for water. There are people who have said that the ostrich never drinks.

However that may be, his appetite is the most curious part of him. He will swallow almost anything he can pick up, and you might wonder where he did pick up the things that have been found in his stomach, were it not for the caravans that now and then come across the desert. Pieces of leather, nails, lumps of brass or iron, to say nothing of stones, all go down his throat with ease.

He has a huge crop, and then a great strong gizzard. And besides these, he has a cavity that might be called a third stomach. So he is well provided.

Of course, strong as his digestion may be, he cannot digest either nails or stones; and some people explain this by saying that his great crop wants so much to fill it, that he is obliged to put in all he can get. And others say that the stones and brass and leather help him to digest his other food, in the same way that grit or gravel helps our poultry at home.

The next curious thing about the ostrich is the pair of wings that Nature has given him. The wing is Nature's machine by which the bird can support itself in the air, and dart or sail through it, as we may see every day. But in some birds the wing fails of this purpose; nay, is of no use at all to fly with. There are two reasons why the wings of the ostrich cannot bear him into the air. They are very small to begin with, and his great body is too heavy to be raised by any such means. And besides this, the feathers of the wings are different to those of other birds.

Look how firm and compact is the wing of the swallow or the rook. The feathers fit close together, and the little plumes on each feather hook into each other by those exquisite little catches that are among the marvels of Nature. If you pass your finger over the wing it feels like one smooth surface. But in the wing of the ostrich the little plumes are loose, and float lightly about. The ostrich does not use his wings to fly with, though he spreads them out as he runs.

He is in many respects so like an animal, that he forms almost a link between the animals and the birds. Indeed, he is so like the camel that he is called the camel-bird. His foot resembles the hoof of the camel. It has only two toes, and both point forward; and the first is longer than the second, and ends in a thick hoof-like claw.

And the habits of the ostrich resemble those of the camel; they both live in the sandy desert, and are able to go a very long time without drinking.

The ostrich does not make any nest, but merely scoops out a hole in the sand. When the proper season comes, the mother ostrich begins to lay her eggs; she lays about a dozen, and they are very large, and of a dirty white colour. In the day-time she leaves them under the burning rays of the sun; but when night comes, and the air is cooler, she broods over them.

The natives of the country go out looking for the eggs of the ostrich. One monster egg has in it as much as thirty of our hen's eggs, and is con-

THE OSTRICH.

sidered a great dainty. But they are very careful how they set about the task of robbing the nest. They choose the time when the mother ostrich is away, and then they take a long stick and push the eggs out of the hole. If they touched any of them with their fingers, the ostrich would find it out in a minute, and go into a great rage. She would break all the eggs that were left with her hoof-like feet, and never lay in that place again. Sometimes a number of mother ostriches will lay their eggs in the same nest.

The flesh of such a great bird as the ostrich is, as you may think, not very tender. It was considered unclean by the Jews, and the Arabs, for the same reason, will not touch it. But when Rome was at the height of her luxury, people hardly knew what fresh dishes to invent, and a dish of ostriches' brains was as great a luxury, and more difficult to get, than peacocks' tongues. A gluttonous and cruel emperor had as many as six hundred ostriches killed to make one dish! And we are told of another emperor a story that we can hardly believe. It is said that he ate a whole ostrich—cooked, we may be sure, in a very delicate manner—at one meal!

In some parts of Africa there are tribes of men who are glad to eat the ostriches, not from gluttony, but because they can get very little else. They keep them as we do cattle, and make them quite tame. The ostrich is by nature gentle, though it is so large, and soon makes himself contented near the dwelling of his master. Sometimes his master rides upon him, and takes a journey, not on his camel, but on his ostrich.

A traveller was once staying in a village where there were two tame ostriches. Two little boys used to mount on the back of one of them and have a ride. The ostrich would run round and round the village, and never seemed inclined to stop.

At first his pace was a trot, but, by degrees, he expanded his wings and ran very fast indeed, scarcely seeming to touch the ground. No race-horse in England could have kept up with him, though the ostrich would have got tired very much the soonest.

The beautiful feathers of the ostrich are so admired, that great pains and trouble are taken to procure them.

The Arab comes with his swift horse in search of the ostriches.

A flock of them are quietly feeding together on the plain. If it is mid-day, they strut about, fanning their wings as if for coolness.

When they perceive the enemy they begin to run, at first gently, for he keeps at a distance, and does not wish to alarm them more than he can help.

OSTRICH HUNT.

The wings of the bird keep working like two sails, and he gets over the ground so fast that he would soon be out of sight if he ran in a straight line.

But he is so foolish as to keep running from one side to the other. The hunter, meanwhile, rides straight on, and when his horse is knocked up, another hunter takes up the game, and so on, allowing the poor bird no rest. Sometimes, in a fit of despair, he hides his head in the sand.

And ostriches have even been seen to swim, a fact not generally known.

THE AMERICAN OSTRICH.

One more thing we might tell you before we bid good-bye to the ostrich.

If he chances to live near a cultivated spot, he is not a very pleasant neighbour. He will go striding on his long legs into the fields of grain, and pick it all out, leaving nothing but the bare stalk. The farmer is very much enraged, and goes out with his gun, to see if he can get a shot at him. But the ostrich is very cunning. He bends down his neck as he eats so that he cannot be seen, and generally contrives to get away in safety.

THE EMU.

THE ostrich has a very near relation, called the emu, that lives in New South Wales. He is larger than the ostrich, and has three toes instead of two. But the toes, as in the ostrich, all point forward.

He may be said to have no wings at all, for unless they are lifted up no one can see them, and they look more like rough hair than plumage. The colour is a dull brown, mottled with a dirty grey. On the head and neck the feathers are still more like hairs, and are so thinly scattered on the throat and ears that the purple hue of the skin is clearly seen. And there is a parting down the back where the feathers divide and fall on either side.

There are many of these curious birds in New Holland, and they are hunted in the same manner as the kangaroo, by men and dogs.

We should tell you that the dogs, as a rule, very much dislike meddling with the emu, and have to be trained before they are of any use. In some places they are taught to go into the woods and look out for the game, and to come back to their master's dwelling, and make known to him the spot where the emu or the kangaroo is to be found.

Then the hunters follow them. The native thinks more of the flesh of the emu than he does of the kangaroo, and he gets very much excited, and he and his companions set up shout after shout, that echoes far and wide.

The emu runs very fast indeed, and the swiftest dogs have great difficulty in overtaking him. When he is overtaken, he stands, like a stag, at bay, and often kicks out his foot, and so wounds the dogs. But the well-trained dog is taught to spring at the emu's neck, and keep out of the way of his foot.

But when the hunt has been successful, and the bird has been killed, the banquet that follows is a very select one. Only a favoured few are allowed to partake, and the young men are not permitted to touch it; if they do, they are severely punished.

The flesh is said by a traveller to be very delicious, and something between that of a turkey and a sucking-pig. But, at the time he partook of it, he and his companions were in a state of famine, which might account for his liking it so well.

The English settlers say it is a little like beef, both to look at and to

THE EMU.

taste, and that the flesh of the young ones is very tender and delicate. The hinder quarters are the only parts that are eaten, and they are such a weight it is no easy matter for a man to carry them home on his shoulders. The eggs are about the size of those of the ostrich, and at the proper season the natives almost live upon them. They are of a beautiful dark green colour, with a rough surface like the coarse rind of an orange, and are laid in a hole like those of the ostrich.

The emus wander about in flocks, and are not very shy. A party of travellers once met with a flock. There were as many as thirty-nine together. They came to stare at the travellers' horses, and were so much interested in looking at them that they did not seem to notice the riders.

The emu has a hollow drumming sort of note.

THE BUSTARD.

THERE is a bird, now very rarely seen in England, that belongs to the same family as the ostrich and the emu.

It may be said to be the representative of them in our own country.

The bird we speak of is called the bustard. Like the ostrich, it has no hind toe, and the legs are long and strong. Its plumage is full and compact, and the wings of moderate length and breadth. They are not useless to the bird, like those of the ostrich, but it can only rise very slowly from the ground, and takes some time before it can gather air enough to leave the ground.

The bustard is much larger than the turkey, and its flesh is so delicate that it would be eagerly sought after were it at all plentiful. But it is a bird that loves the open plain, where it can see all round; and while there were such places left in England, it lived in flocks. But now that the country is covered with meadows and corn-fields, there are no retreats left for the bustard. It is very rarely seen, except on wild and solitary commons, such as Salisbury Plain, or the heaths of Sussex and Cambridgeshire, or even in the wild districts of Scotland.

The flat plains of Norfolk, called "the bustard country," form an excellent home for the bustard, and about fifty years ago it used to be abundant there. And about the city of Norwich an attempt was made to keep bustards in a domestic state like the famous Norfolk turkeys. Their flesh was so delicious, and so much sought after, that it was wished to increase the

GREAT BUSTARDS.

numbers. By good care and feeding, the farmer's wife hoped the birds would lay eggs, and rear the broods, but this was not the case.

For the last half century scarce a bustard has been seen, and every time such has been the case, the fact has been recorded with interest.

One large bird was taken on Newmarket Heath, and sold in London for five guineas.

A gentleman, who was a great sportsman, once declared that he saw

THE LITTLE BUSTARD.

a flock of bustards rise before his gun, in a sudden manner, out of a gravel pit; and it was even reported that the bustard had been known to attack a man on horseback at night.

Once a solitary bustard haunted a turnip-field in Cornwall, and the country people fancied it must be an eagle, from its great size and the noise it made when it rose from its covert in the bush. In Bristol there is a record that one Christmas two bustards were served up at table, and gave great magnificence to the feast.

Another story, not so probable, is that the last of the Salisbury Plain bustards came up to a farmer, and let itself be taken as if in despair.

The bustard feeds on the berries that grow on the heath, and on the large worms that lie in quantities on the downs before the sun rises on a summer morning. Its food is moist, and it can live a long time without drinking. Indeed, one that was kept in a tame state never drank at all.

The male bird has a curious pouch or sac that opens under the tongue, and can hold three quarts of water. People have thought that the water was for the use of his partner when she sat on her eggs; but this is not the case, for it is said that as soon as she begins to sit he deserts her, and does not return till the young are fledged.

The bustard runs with all the speed of its race, and you might think would easily get out of the way of the dogs. Indeed, it can go several miles without stopping. But its danger is from attempting to fly. It flaps its wings, and endeavours to get the air beneath them in order to rise; but much time is lost in this way, and the dogs get nearer and nearer. Then it has to give up the attempt and keep on running, until it either gets away or, if it is too fat to run as fast and as far as usual, it is taken.

The nest is made upon the ground, by scraping out the earth after the fashion of the ostrich, and sometimes lining it with a little straw.

Two eggs only are laid in the nest, about the size of a goose's egg, of a pale olive brown, and marked with darker spots. The young are hatched in five weeks, and run about as soon as they leave the shell.

In October the bustards assemble in flocks, and keep together until April. In the winter their food becomes scarce, and they catch all they can find; moles, mice, and even little birds become their prey. And, for want of other food, they live on turnip tops and what vegetables are to be had. Sometimes, in very severe weather, they are frozen to death; but the natural term of the bustard's life is fifteen years.

In some parts of the continent the young are taken alive and kept in confinement. They are fed with rye bread, mixed with the yolk of an egg; and then with rye bread chopped up with bullock's liver.

In France the bustards are considered birds of passage. They come about the beginning of December, and assemble in small groups of thirty or forty, and betake themselves to the vast plains of Champagne and Poitou. If the winter is very severe, they are more widely scattered, but they prefer those spots which are remote from human habitations, and where they can see to a great distance.

All kinds of stratagems are used to catch them. The peasant disguises himself in the hide of a cow, and creeps on all fours, in order to deceive the birds, or he covers himself with a little hut, and steals along, fancying he shall not be seen.

In the Crimean War the British officers amused themselves with hunting the bustards which inhabit the central plains of Asia.

The flesh of the bustard is held in great esteem on the continent, and is exposed for sale in the markets.

THE PRAIRIE HEN.

THE proper name of the very curious and beautiful bird in the picture is the pinnated grouse, and it belongs to the grouse family; but its popular name is the prairie hen.

There is a great deal to say about the heath hen, and we think we must first take you to a few of the places where it lives.

The barrens of Kentucky are not, as their name implies, bare and sterile, but in their season teem with beauty. Here are vast plains covered with flowers and verdure; and here, also, are the orchards and the fields of the settler and his homestead. And so luxuriant is nature that the wild fruit-trees have their branches interlaced with the vine, and ripe strawberries carpet the ground. Here, too, are groves and valleys, and springs of clear cool water. And here is the home of many a living creature that dwells in security and plenty. The graceful deer glides along, and the wild turkey leads out her brood amid the grass herbage.

Here, too, there used to be heard very often a curious booming sound, as though the voice of some person at a distance had come strangely to your ear, or as though a horn were being blown a mile or two away.

If you had searched further into the matter, you would have found out that the sound proceeded from a large bird, with two great bags of yellow skin on each side of its neck, which sometimes hang in wrinkled folds, but when he is making the noise of which we are speaking they are full of air, and swell out to the size of an orange.

The sound consists of three notes, the last being twice as long as the others. The male bird utters it to his mate, and as he does so he struts about and flutters his wings in the same way as the turkey cock does. And sometimes he makes a cackling noise that is quite laughable, and that is meant to defy the other birds, his companions, who boom and cackle in return, and often fight each other.

And at this season the booming will keep on from daylight to about nine o'clock, when the birds leave off and disperse to find their breakfasts.

We must give you a little description of the prairie hen before we go any farther.

It is, as you see, extremely handsome; and on its neck are two little wings, besides the larger ones, of a brown black colour.

There is a small crest on the head, and over the eye is a comb of a beautiful orange colour, which the bird can set up or down as it likes. Its chin is cream colour, and the upper parts are mottled with black, brown, and white. Its feet are covered with hairy down to the toes.

The mother bird is a much less size, and has not the extra pair of wings, nor the yellow skin on the neck; nor has she the handsome comb over the eye.

We should tell you that the place where the male birds strut about and boom and swell out the bags on their necks is called a scratching place, and is well known to the Indian. He makes a little hut of pine branches, and remains snugly hidden until the performance begins. First a low booming is heard, and out steps a cock bird from the bush, and begins to strut about. The boom is answered from another bush, and another cock comes forth and swells his neck and sets up the wings on it like a ruff; the plumes of his tail expand like those of the turkey cock, and he eyes his fellow with anger and disdain. Another, and then another, comes forward, and the whole party strut and boom and provoke each other, until fierce battles begin, and they scream and leap from the ground, and grow very excited indeed.

The Indian, from his lair, sees all this, and watches the moment when two cocks are fighting with fury. He has his gun in his hand, and fires with

THE PRAIRIE HEN.

effect. Indeed, he has killed so many birds in this manner, that they begin to be afraid of coming, and will perch on the trees instead of alighting on the ground.

There was a time when the pinnated grouse was very plentiful indeed in that part of the world. The hunter of Kentucky hardly took the trouble to shoot them; and in some places they did a great deal of mischief among the fruit-trees in the orchards, and picked off the young buds; and also among the grain in the fields.

The farmer used to set his children to frighten them away with rattles from morning till night; and he set traps for them, and did all he could to destroy them. In the winter they were so tame they would come into the farmyard and feed with the poultry, or even walk in the village street like the cocks and the hens. And when they were shot they were often left to lie dead on the ground. The hunter did not care to pick them up, for he had eaten so much grouse that he was more than satisfied.

But times have altered strangely since then. There is scarcely a grouse to be seen now, so have they been hunted down. Farther and farther have the little remnant retired into spots where they can escape the hunter, and game laws have been made for their protection. The sportsman now travels far and wide with his dogs to get the chance of a shot. And in the markets of New York and the large cities a brace of these birds sells for one or two sovereigns.

The open ground of the prairie is the place chosen by the mother bird to make her nest. It is carelessly built of dry leaves and grass woven together, and is placed in a large tuft of grass or at the foot of a small bush.

There are seldom more than twelve eggs, and as soon as the young ones are hatched the mother leads them away, and is deserted by her partner.

The little birds enjoy themselves in the boundless home of the prairie, and find an ample store of food. They squat so close to the ground that it is hardly possible to see them. When there is the least danger the mother gives an expressive cluck. The little birds know what it means, and do what she wishes them. They spring up for a moment on the wing, so as to confuse the eye of the enemy, and then lie down quite flat. Nor is he able, with all his searching, to find out where they are, although, at the same time, he may nearly tread some of them under foot. While the young ones are skulking in this manner, the mother bird does all she can to entice away the intruder from the spot. She droops her wing as if she were lame, and limps about, and even rolls over on the ground, and diverts his attention until he is at a safe distance from the brood.

They have other enemies besides the sportsman. The owl, the hawk, and a disagreeable small animal called the skunk, prey upon them when they are young and feeble. In winter their stock of berries gets less, and they fly to the tops of the trees to feed on the seeds. As many as fifty have been counted on the boughs of an apple-tree, and they destroyed all the buds in a few hours.

They also alight on the high forest trees near the great rivers, and fly across the mighty streams in flocks.

THE PTARMIGAN.

THE smallest of all the British grouse is the ptarmigan. It once had a home in the mountain ridges of Cumberland, but is now only seen in the hilly parts and the isles of Scotland.

Its chief home is in the mountainous parts of Europe, among the bold scenery of Norway, choosing, however, as a rule, the districts that have no trees. But in heavy snow-storms the birds will descend to the valley, and even perch upon the birch-trees, till the branches appear as if clothed with white.

There is another species of ptarmigan in Norway that is of a larger size, and is called the yellow ptarmigan.

The great naturalist Linnæus met with both species in Lapland. In one of his excursions he came upon the lesser hen bird, surrounded by her brood, and he picked one of the little ones up. Upon this the mother bird was in great distress, and came so close to him that he could have taken her as well. She jumped round and round him, as if asking him to give back her young one, until at length his heart relented, and he restored it to her in safety.

The ptarmigan has a black beak, and a small black patch behind the eye. Almost all the plumage is pure white; but some of the tail feathers are black, tipped with white. The legs and toes are white and the claws black. This is his winter dress. In summer the head and neck are mottled with speckled grey feathers, and the breast, back, and upper tail feathers speckled grey. The hen bird is smaller than her mate, and is pure white in the winter;

THE WILLOW PTARMIGAN.

but in the spring her costume changes, and the feathers have a mixture of black and yellow, with white tips. In Norway the ptarmigan is taken in snares that are set in the snow. A forked twig of birch is stuck into the snow, so as to make a kind of arch. A noose made of hair is fixed to the twig between the forks, and the traps are set very near together, just in the way where the ptarmigans come running along, for they very seldom fly. They run into the snares by as many as forty or fifty at a time.

THE PTARMIGAN IN WINTER PLUMAGE.

Sometimes the hair noose is found round the neck of the bird when it comes to the London market.

One peasant will set as many as a thousand of these snares.

THE QUAIL.

WHEN the children of Israel were wandering in the wilderness, and were fed by manna from heaven, they murmured for flesh to eat; and we are told that quails were sent in such quantities that "feathered fowls were like the sands of the sea. So they did eat, and were well filled: for He gave them their own desire."

There has been a great deal written by learned men about this miraculous flight of quails, and to what species they belonged. And it is interesting to know that the bird in the picture is thought to be of the very same family. It is the only species of quail that ever takes long journeys, or flies in large flocks. And old writers tell many stories of the vast quantities of quails that were sometimes seen flying from place to place. Indeed, one of these writers declares that the birds sometimes settled on a ship in such numbers as to sink her!

Whether this be true or not—and it sounds rather like a fable—the quails have been seen in our own days in certain countries flying in countless numbers. At the proper season, the islands in the Archipelago are covered with them. And the bishop of one of these islands, near Naples, derived the chief portion of his income from the quantities of birds that were caught on his island, and he was actually called the "Bishop of Quails."

And on the western coasts of the kingdom of Naples as many as a hundred thousand are taken in one day.

These monster gatherings of the quails are in the spring, when they pass to the northern parts of Europe, and in the autumn, when they go southward. They arrive from Africa in thousands, we might say millions, in the month of April, and spread over Europe, touching even at our own country.

And when we think over these facts, the vast numbers of the birds, their habits as birds of passage, and also their custom of flying by night, we cannot but recall the words of Scripture, and apply them to this very bird: "And it came to pass at even that the quails came up and covered the camp."

The quails reach our shores in May, and betake themselves to the open country. The male bird arrives before the female, and in France the bird-catcher takes advantage of this. He goes out and imitates the note of the

THE QUAIL.

hen bird, so as to decoy a great many quails into his net. Then he brings them to the London market, and sells them for the table, their flesh being thought a delicacy.

The mother bird scrapes a little hole in the ground, and fills it with grass, or stalks, or clover, and lays her eggs upon it. They are of a dull orange colour, speckled with brown.

THE QUAIL.

The quail is much smaller than the partridge, and the feathers of its head are black, edged with rusty brown. The breast is a pale yellow red, spotted with black, and the feathers on the back are marked with pale yellow lines.

The young birds soon learn to follow their mother, and feed on seeds, grains, and insects. They are often killed by the sportsman in the stubble fields in the autumn. But in October most of them leave England, though some remain and are seen during the winter.

THE CURLEW.

THE curlew belongs to the family of birds called waders, from their habit of wading into the water in search of insects. They are not formed by nature for either swimming or diving, as their legs are too long to propel them in the water, and they have no web foot. Yet, now and then, the curlew has been known not only to swim, but to dive, though it is not provided, as the divers are, with an air-bag in the throat.

We can tell you a little story about a curlew. It had been wounded in the wing by a gun, and fell into the sea at some little distance from the shore. The sportsman who shot it had no idea it could swim, and he had not his dog with him to fetch it out. He threw off his coat and jumped in to fetch it himself; but when he came near the bird it began to swim away from him. He was not able to catch it, and as it went on swimming, it led him a long way from the land; in fact, he was obliged to give up the pursuit, and go back again.

The wading birds have all very long bills. Their food lies underground in the soft mud, and they have to feel for it. So that Nature has given many of them a bill that can feel, and can easily be driven into the earth. The bill of the curlew, and some other of the waders, has six large nerves passing along the roof of the mouth, and extending to the point of the beak. The whole beak is therefore sensible both to taste and to feeling. Their feet have often to stand on slippery places by the water's edge, so that they are wide-spreading, and have long toes. The legs are generally tall, to raise the bird out of the mud.

The wading birds only frequent the sea-coast during the winter. In the warm days of spring they fly away to the north, and seek the fens and moors far from the abodes of man. The sportsman who wishes to see them in their summer haunts must prepare for many difficulties, and think nothing of a walk through a quagmire of reeds and rushes. And if he makes the slightest noise, and does not creep along in the most cautious manner, the birds will be gone before he catches sight of them. When on the moors they are scattered about, but on the sea-shore they form themselves into flocks, and may be seen following each other in a long line as they wade about on the sand, picking up crabs and worms. The mother bird is so like her mate that she can hardly be distinguished from him.

The curlew is a very common bird, and well known on all our wastes. Its home and haunts are amid the wildest spots in Britain, and to glance

THE CURLEW.

at it in its every-day life is to give a sketch of some wild waste or sandy pasture, near the sea, and that is sometimes covered with the tide, and then left bare.

Here are gulls flying about with their silvery plumage, and cormorants far out at sea, busily catching fish, and golden plovers, and snipes, and hundreds more, all enjoying the loneliness and the security of the spot. Among them stalks the curlew. You see that he has long legs, like the heron, and a long bill. He thrusts his bill into the soft sand, and works it about. Then he draws out a worm, which he swallows, withdraws his bill, and looks carefully round. By-and-by he picks up a snail, and now and then he discovers a crab lurking behind a stone.

He is very shy and suspicious, so that it is said by the sportsmen in the Hebrides, that to kill seven curlews is enough for a lifetime.

When alarmed they spread out their wings, run forward a few yards, utter loud cries, and fly away.

They equally frequent dry pastures and moist meadows and shallow pools, and are often seen wading into the water.

Towards the end of March they leave the sea-shores, and betake themselves to the heaths and moors, and begin to make their nests.

The young curlews soon begin to run about, and are covered with long stiff down. They squat down, if they are alarmed, quite close to the ground. At first the bill is only about the length of the head, but it soon grows longer. At this period both birds, old and young, feed upon insects, larvæ, and worms. But they do not get fat until the autumn, when they unite into small flocks, and go to the coasts. The flesh of the curlew is thought to be a great delicacy, and the birds are often met with in the markets.

We can tell you a story of a tame curlew. He had been wounded in the wing, and was carried home and put with a number of other birds of the water-loving kind, such as ducks and geese. He was so very shy that he would not eat anything, and would have been starved to death if his owner had not crammed the food down his throat. After that, he grew tamer, and began to eat worms. As worms were not very plentiful, an attempt was made to feed him with bread and milk, and a few worms were put into the mess to tempt him. It was curious to see how he acted. He would pick out a worm, having carefully hunted for it with his bill, and then he would carry it to the pond, and carefully wash it from every particle of bread and milk, before he would swallow it. After a time, however, he began to like the milk and bread, and would eat heartily, and from being skin and bone he grew plump and in good condition.

He also became very tame, and would follow any one about for a bit of bread or fish. He began to eat almost anything—fish, frogs, or insects; and if nothing else was to be had, he would eat barley with the ducks.

He became a very great favourite, but was at last killed by a rat, after having been two years in captivity.

The curlew is found on all the high moors and heaths from one end of the kingdom to the other, but nests are more numerous in Scotland than in England. In Turkey and Holland they are found all the year round.

It is thought that the birds that pass the winter in the south of England spend the summer on the Grampian Hills. But their comings and goings may be merely from the hills to the sea-shore and back again, after the custom of the plovers and the lapwings.

The family of curlews are dispersed all over Europe and Asia, and in Africa extend to the Cape of Good Hope, but they are not known in America.

THE PLOVER.

PLOVERS have their home on dry and sandy plains or heaths, or on the sea-shore where it is lonely and unsheltered.

Their feet are long and slender, and adapted for running very fast indeed. The toes are short, and the hind toe is entirely wanting. The head is thick, with large dark eyes placed rather far back; and the bill is about the length of the head, the outer half slightly notched, a little like that of the pigeon.

The most beautiful of all the tribe is the bird in the picture, that is called the "golden plover," and has a costume of brownish black, spotted round with yellow. The wings are a chocolate brown; and the fore part of the breast is black grey, bordered with white.

The golden plovers live in every part of Britain during the winter. In open weather they are scattered about on plains or ploughed fields. But when the frost seals up the ground, they betake themselves to the beach, and run about on the sand, picking up what they can find, and now and then wade a little into the sea. Thus, for a season, they contrive to gain a livelihood. But when spring comes, they seem to remember their native moors and commons, and then they gather in straggling flocks, and fly away in that

THE GOLDEN PLOVER.

direction. As they fly they utter soft and pleasing notes, mingled with which comes a curious cry, like the word "counter-wee."

When they have reached their wild haunts, the male birds appear clad in their showy summer costume, and are very tame. Often, as the tourist looks about him, the bird comes and alights close by on some mossy bank. This is not the season for the sportsman, so he escapes with impunity.

The nest is a very slovenly fabric, and consists of a few fragments of heather plants, placed in a slight hollow, or laid on a dry spot on the ground. There are four eggs of a light yellow, or cream colour, dotted and patched with brown, and sometimes with a few light purple spots.

The little plovers come from the shell quite able to run, and to take care of themselves, and to leave the nest at once.

The mother bird goes about with them, however, and is very anxious for their safety, and if any danger threatens, is in the greatest distress. She does all she can to distract the attention of the enemy. She flaps her wings and droops them, as if she were too lame to proceed many steps. And she will even flutter about on the ground, as if she were in the agonies of death. All this time the little plovers are lying flat on the ground, so that not a trace can be seen of them. In very wintry weather also, the plovers have the habit of squatting close to the ground.

When the young birds can fly the plovers begin to collect into flocks, ready for their winter journeys. But they remain on the moors as long as possible.

The plover has other enemies besides man. The peregrine falcon has been seen to pursue and seize him. And a battle scene has been viewed between the hawk and the plover.

It lasted nearly ten minutes, and the plover doubled about like the hare before the greyhound, sometimes darting up into the air high above the hawk, then hiding behind a bush or crag. The hawk took matters very coolly, and did not fatigue herself in the least. But she never gave up the pursuit, and kept quietly on until her victim began to be exhausted and slacken his pace. This was the moment the hawk had been waiting for. She made a sudden pounce, caught the bird in her talons, and carried him off.

THE LAPWING.

OFTEN, in some solitary walks over the fields or commons, you may see a number of birds with large wings flying over head, uttering a curious note, like the word "pee-wee," or "pee-o-wee." The wings look larger than they really are, on account of the black colour underneath contrasting with the pure white of the body. And the birds move, and tumble, and glide about in the air in the most spirited manner, flapping their wings with violence.

They perform this feat for no particular reason, but simply to amuse themselves while their mates are sitting at home and hatching the eggs. The nest is in some part of a moor or field where the grass is short, and it is slightly built of a few stems put together in a hollow place, and, because of the colour of the eggs, it is very seldom seen. But, should your feet turn in that direction, the mother bird will spy you out, even at a great distance. She will rise up and approach you, flying about in a state of excitement, and trying to lead you from the nest. And the lapwings in the neighbourhood, as if quite understanding the matter, will come and join her, and fly, and flap, and "pee-wee" over your head with great energy.

All at once, however, it appears as if the mother lapwing had suddenly become lame. She runs limping along, and it seems the easiest thing on earth to catch her. She will allow you to come very near indeed, and entice you to a great distance. Then, when all danger is over, she will spring up, as if laughing in your face, and fly off.

The lapwing has, as you see, a beautiful crest of black feathers upon his head. His neck and throat are a deep rich black, with a gloss of green. The sides of the head and neck are white, with a black shade under the eye. The upper parts of the body are a pale brown, with shades of blue and purple, and the under parts are pure white, while the tail is black.

The lapwing is larger than the plover, and differs from it in having a minute hind toe. Like the plovers, the lapwings migrate in the winter to the sea-side, and numbers of them leave us altogether, and return in the spring.

Large downs and sheep walks, heaths, pastures, and rather wet meadow land, are the favourite haunts of the birds. At the season when their family cares begin, the moors seem alive with them.

THE LAPWING.

If any stranger approach they fly over his head, and tumble about in a state of excitement, uttering their loud cry without ceasing.

When the little pee-wits are old enough, both young and old assemble

THE LAPWING, OR PEE-WIT.

together and fly to the fields and pastures. They feed principally in the night, and rest in the day-time.

About sunset a beautiful sight is often witnessed in these spots. A cloud of birds, that have been resting all day, rise up in the air. In the flat country of Holland, where there are millions of lapwings, this sight is seen to perfection. Thousands of birds on all sides gleam in the setting sun, and look like a dense mass, almost hiding its light.

The eggs of the lapwing are thought to be a delicacy, and are seen in the London shops in great numbers in the spring.

THE SPUR-WINGED LAPWING.

Collecting the eggs is a trade carried on by many persons during the season. The low flat counties near the metropolis are well searched. Great expertness is shown by those who are accustomed to the business. The mother lapwing, as we have seen, does all she can to mislead them. But they quite understand her manœuvres, and go just in the direction she is trying to prevent. Her partner wheels round the heads of the intruders, and makes a clamorous noise, as if to frighten them away.

But the egg-collector will walk straight to the nest, not at all alarmed by this display of hostility. Sometimes dogs are trained to find the eggs.

The food of the lapwing consists of worms, insects, and grubs. It has a very cunning way of enticing out the worm. It picks down the worm-hill with its bill, and then walks round it. The worm comes out to see what is the matter, and is instantly seized and eaten.

The lapwing is a very useful bird in a garden, on account of its habit of devouring insects.

THE WOODCOCK.

SOMETIMES, on a still October night, when the wind is in the north-east, a great cloud of birds comes silently over from their summer quarters in Norway and Sweden, to spend the winter in England. They are well known to the sportsman by the name of woodcocks, and he does not give them a very kindly reception.

On the contrary, if he is acquainted with the spot where they have alighted, he takes his gun and goes out, feeling sure of an excellent day's shooting. He is almost certain to find them in great numbers in hedges by the side of plantations, or even in turnip-fields; and he kills many of them, making the most of his time. The next day they may have gone away—having resumed their flight as soon as it was dark.

They are not at all adapted for a long flight, and can only accomplish it by choosing a favourable wind. A captain of a ship plying to and from Norway and Sweden declared that he had often seen them, when quite spent and tired, drop a moment or two on the smooth water behind the ship, and rest with outspread wings. Then they seemed revived, and continued their voyage.

The bird belongs to the same great division as the heron and the stork; but the smaller family to which it belongs differs from them with regard to the beak. The beak is not hard and firm, like the heron, but soft and flexible. The tip is covered with a soft skin that is very sensitive, and the bird uses it to thrust into soft earth, and catch little insects or worms that it could not otherwise find. The hind toe is placed higher up than the front ones, and does not reach the ground. In some birds of the family it is quite absent.

The favourite food of the woodcock is earth-worms, and it seeks after them with the greatest eagerness. The quantity it will devour is almost past belief. There was a pair of woodcocks kept in an aviary in Spain, and every care was taken of them. It was a delightful spot, with a fountain trickling on the ground to keep it moist, and shaded by trees from the heat of the sun, and fresh sods were constantly supplied. The woodcocks had an ample supply of worms, and it was an amusement to their owners to watch the process of catching them.

In vain the worm buried itself in the earth; as soon as the bird was hungry it was sure to come to the exact spot. It seemed to find out its prey by the smell, and never missed its aim. It would plunge its bill into the earth as high as the nostrils, and draw the worm slowly out; then it would raise its bill in the air, and swallow the prey in an instant.

The plumage of the woodcock is very compact and variegated with different colours, such as black, brown, grey, and yellow. The forehead is grey, and there are three pale red bands on the upper and back part of the head.

It is a bird of night, and hides during the day in some secure retreat under a bush or a tree; and the sportsman has to beat about with dogs in order to start it. But when it is not disturbed it does not always doze all day, but has been seen searching actively for worms and insects along brooks and ditches, and by the side of hedges. A laurel or holly bush is a favourite spot for the woodcock to repose under, the thick leaves affording a warm shelter from the cold.

Towards night the bird issues forth on silent wing, and takes the usual track to its feeding ground. The glades in the wood through which it passes are well known to its enemies, and are called cock-roads. At one time nets and snares were laid in its way; but now the more usual mode of procuring it is by the gun.

In Scotland the fir woods form a pleasant shelter for the woodcocks in the day-time; but if the weather is very severe, they quit their lair in the wood during the day, and go out to feed at the sheltered places where the frost has relaxed its hold on the ground.

The bird drives his bill in the mud to a great depth, and performs the operation of plunging and drawing back so fast that you can hardly keep count.

When he sees you he stops, draws his bill into his breast, and squats upon the ground, and becomes quite motionless. Indeed, you can hardly distinguish him from the ferns and withered leaves around him.

THE WOODCOCK.

The woodcocks begin to make their nest very early in the year. At one time it was supposed that they did not remain in England after the end of the winter; but latterly many of them have done so, and reared their young. The reason is supposed to be because of the increase of fir and pine plantations in many parts of Scotland.

The nest is often placed on the ground, at the foot of the Scotch fir, and is made of the leaves. There are two broods a year.

The mother woodcock is very devoted to her young, and will carry it in her claws from any threatened danger. She has been seen to fly over a road, carrying her little one with her in this manner.

Though some of the woodcocks remain in Britain during the summer, yet by far the greater number leave us in the spring, and return to their old haunts in the north of Europe.

THE HERON.

THE heron, as he stands fishing for his food, is the very picture of patience. For some time he has been slowly stalking about on his long legs, watching for his prey.

Now he approaches stealthily into the water, laying down one foot after another with the utmost caution. He does not want to alarm the fish that may be swimming merrily about, or the little fry that is enjoying itself in among the stones. He does not want either of them to know that he is there.

His custom is to stand on one leg, his neck drawn up, his eye intently fixed on the water below. He will stand thus for hours, until you hardly know whether he is alive.

Suddenly, however, the happy moment comes. The fish that had perhaps caught sight of him and swam away, has forgotten his fright; and the fry that lay hidden under the stones feel sure he must have gone by this time. But they do not understand the nature of their enemy. Nothing can weary out his patience or make him forget.

The moment the victim comes within the right distance, down goes the bill with its sharp edge, and the prey is seized and devoured.

He has an excellent appetite, and can devour more in a meal than you would believe. When he has finished eating he goes away into some quiet place, and stands on one leg for hours.

He may be called the prince of the wading birds; and the heron in the picture is the largest of his tribe. He is very rarely seen in England, and,

THE GIANT HERON.

indeed, the common heron is by no means so plentiful as he used to be, while the great white heron belongs to America.

He is dressed at all times in a costume of spotless white, slightly tinged with cream colour. His flight is firm and regular, and can be kept up a very long time. He urges himself forward by slow regular flaps, his head drawn in, and his legs extended. Now and then he sails in wide circles, and when he alights he always wheels round and round before he settles.

When a white row of herons are standing watching for fish along the river, they look, against the blue sky, like statues of the purest alabaster. They do not move towards their prey, but watch till it approaches them;

THE PEACOCK HERON, OR SUN BITTERN.

they swallow it alive, or, if it is large, beat it on the water or shake it until it is dead. They do not move until the tide drives them back, and even then remain until the water reaches their body.

They roost on trees, but rarely alight on the same tree, as if they were afraid of being discovered by their great enemy, man.

Often, while the flock are asleep during the day, two or more herons stand, with outstretched necks, acting the part of sentinels. If there is a splash in the water, as some great fish, such as the shark, gives chase to another fish, the heron sentinels give a start, and seem much alarmed.

GROUP OF HERONS.

Or if there comes the sound of oars, and a boat comes down the stream, it causes the utmost terror. Yet few people are known to molest them.

They place their nests among the tall reeds, at some distance one from the other, and only a few feet above the high water mark. The nest is large, and made of sticks, without any lining, and is quite flat. The eggs have rather a thick shell, and are of a light blue green colour. Both birds sit on the eggs, which take a month to hatch.

Before we leave this bird we must say a few words about the night herons that live in the cedar swamps.

The cedar swamp is perhaps the most dismal spot you can imagine. The ground underfoot is like a bog, covered with great bushy limbs and logs of fallen trees. And the trunks of the cedars grow side by side to the height of two hundred feet, and so close together that a man cannot push himself between. And there are no branches, except at the top, where the trees are all matted together, so as to shut out daylight.

Nothing breaks the dreary silence except the chirp of a few birds, or the harsh screams of the heron. And if the wind gets up, the tall stems clash together, and rub one against the other, and make such creaking, and such hideous noises, that the effect is something awful.

Every spring the herons come to the cedar swamp, and take possession of their old nests on the cedar-trees. All the branches near the place where they live are completely battered and broken down by them, and the ground is strewed with feathers, and fishes, and pieces of old nests, and all kinds of rubbish. And we can hardly describe the noise, for it is enough to deafen you. They keep repeating the note "qua-qua," until the Indian gives them the name of "qua-birds."

The little herons are some time before they know how to fly, but they soon begin to crawl about the branches, and get to the top of the tree to look out for their parents. They are terribly afraid of being caught, and if by chance any one comes that way—and the Indian thinks young heron as nice as pigeon—they scramble out of the way as fast as they can, and hide themselves among the mud.

The common heron is the most familiar of its tribe. Its general colour is ashy grey with a bluish tinge. The blue tinge is deeper on the back of the head, that is ornamented with a crest of narrow black feathers, shading the back of the neck.

THE HERON.

THE GREAT WHITE HERON.

The upper part of the neck is of a light grey, and the wings have a reddish tinge. The under part of the plumage is pure white, marked on the front of the breast with large black spots.

The herons are birds of passage, and their going and coming depend on the supply of food they can obtain. They are nowhere very abundant; but are met with in almost every part of the northern and temperate regions of the old world.

They build their nests in companies, like the rooks, in lofty trees, in the neighbourhood of streams and rivers, and such places are called heronries. They are very fond of the society of the ravens, although the raven often returns their friendship by carrying off their eggs. The falcons and the weasels are also great enemies to the young birds. The heron leaves the care of hatching the brood to his partner, but when this task is over he assists in providing the family with food. When the young birds are strong enough to get their own living, the parents drive them away, and they take each a separate course, and begin the world on their own account.

Their food consists of fresh-water fish, especially of the young of the carp or the trout. The heron is very fond of eels, and catches them in a very dexterous manner. We can tell you a little story to show that in one case he did not proceed with his usual skill and caution.

There are, we should tell you, still many heronries in different parts of England, and in the grounds of noblemen, where some stream meanders through the domain.

In one of these places, at Carlton, in Nottinghamshire, a heron was standing as usual, patiently waiting for his prey, when a fine large eel came in sight. Down went the prong-like bill of the heron, but, in his eagerness, he plunged it too near the head of the eel. The long, snaky body was left at liberty, and it twisted itself round and round the neck of the bird until it strangled him. The heron was found the next day, dead on the bank, with the eel, also dead, twisted round his neck. The owner of the mansion had the two creatures, just as they were, preserved as curiosities, and as such they are still to be seen.

In the winter fish are not so plentiful, and the heron has to be satisfied with frogs and snails and worms, and even the duck-weed that floats upon the pond. At these times he becomes very thin and poor, and is nothing but feathers and bones.

In the old days of falconry, hawking the heron was considered the highest feat that could be accomplished. The powerful wings of the bird enable it to rise so high that it put the powers of the falcon to the test. That was the time when the herons were preserved with the utmost care, and the heronries watched over and provided with every necessary. There are very few of these old heronries left in the country.

The young heron soon becomes tame, and gets reconciled to captivity, but the old ones pine away and die. In the old days, however, and when the heron had to be procured in order to train the hawk to fly at him, he was crammed with food like a turkey. Often, after this had been done, the bird would become tame, and follow his owner about for miles, and come when he was called, and take food from his hand.

In most cases the bill of the fishing bird is lined towards the point with bristles. The bristles point backward, so that the food can slip easily over them, but it cannot come back without being caught on the bristly hooks. There is no crop at all, and the food goes at once into the stomach. The throat of the heron has the power of stretching out when it gulps down a fish too big for it. It stretches into a fan-like shape, and then comes back again when the fish has gone down.

THE STORK.

THE white stork is so seldom seen in Britain, that its habits cannot be studied here. We must visit its favourite haunts abroad to become acquainted with it.

It is a near relation of the heron, and has long and slender legs, and a long and rather thick neck. The bill is as long again as the head, and tapers to a point.

In the countries where it lives it is cherished with the utmost affection. In Holland the people in the towns and cities place wooden boxes or frames on the tops of the houses or chimneys to induce the storks to settle there. And the storks are perfectly tame, and are thought to bring prosperity to the person who entertains them.

The stork goes away in the winter, to Egypt or some warm country, and comes back with the swallows.

o

The ancient Egyptians almost worshipped the stork, and it was one of their sacred birds.

The reason why the stork is so much beloved is because it destroys the snakes, and rats, and mice, and other unpleasant creatures that infest the town. It settles fearlessly on the chimneys and roofs of the houses, and builds a flat nest of sticks, lined with twigs, and straw, and dry grass. There are three or four eggs of a bluish white colour, and it takes thirty days to hatch them.

In Holland and Germany the stork rears her young in the utmost security on the tops of the houses, and even walks about in the most crowded streets, amid men and women and children, without the least danger. To harm a stork is considered an act of barbarity.

The young birds come out of the shell covered with down, and remain in the nest until the end of summer. The parent storks watch over them with the greatest attention, and feed them by putting food into their mouths from their own beaks. Nothing would ever induce the stork to leave her young ones; she would rather remain and perish with them. We can tell you a little story in proof this.

Once there was a great fire in the city of Delfth, in Holland. The flames spread to a house on which a mother stork was rearing her young. The little ones were too weak to fly, and their parents did all they could to carry them away. They made many and desperate efforts, but it was all in vain, and the little ones were obliged to remain in the nest. Meanwhile, the fire came nearer and nearer, and you would think the old storks would be frightened and fly away. But no; they still refused to leave their little ones, and stayed close by them. Even when the flames closed round the nest they did not stir, choosing rather to die with their young ones than desert them; and the whole family perished together.

After such a touching history, one does not wonder that the stork is respected and beloved.

The birds return year after year to their old haunts, and are eagerly welcomed by the owners of the houses. No doubt the useful and friendly birds would attach themselves in the same way to our English homes; but, unfortunately, every stork that has shown her face amongst us has been at once shot down by the remorseless gun, so that there is no chance of their dwelling among us.

THE STORK.

A white stork was seen in Northumberland in the last century, and, of course, shot directly. Its body was nailed to the wall of an inn, and crowds of people came to look at it as a great curiosity.

Since then, as lately as 1830, storks have been seen in England, and four or five haunted a pool on one of the great commons in Yorkshire; but they were not allowed to live in peace, and the little community were soon dispersed.

THE WHITE STORK.

How much might our stock of English birds be increased and improved but for this fatal habit of wanton destruction!

There is a black stork as well as a white stork; and it, too, has now and then been seen in England, and met with the same reception as its relative. In one case a black stork was seen hunting for food by the side of a drain. It was wounded slightly, and taken alive.

The black stork has not the same habits as the white stork. It avoids houses and villages and the dwellings of man, and chooses the most solitary spots in which to make its nest.

The banks of rivers and lakes, or even the top of a tall pine-tree, are among the spots chosen by this unsociable bird. Here the mother stork lays two eggs of a buffy white colour.

The black stork is more delicate in its choice of food than the white one. It is especially fond of eels, and takes them in a very clever manner. No spear used for taking eels can be more effectual than the stork's open bill, that spears them in a minute. But the stork does not swallow the eel all at once. It retires to the margin of the pool, and shakes its prey with its bill before it eats it.

It never swims, but wades into the water up to the body, and will sometimes thrust its head and neck under water after a fish. It has no note of any kind.

The black stork passes the winter in the south of Europe. It is unknown in Holland, where the white stork is so familiar; but is found in Russia and Siberia, where lakes and morasses abound; and occasionally it visits Sweden.

During the whole time of rearing their young the parent storks never lose sight of them. If the mother bird is out on an excursion, to look for serpents and lizards, her partner stays at home to take care of the family. And when the little stork is making its first attempt to fly, the parents keep on either side, and support it in the air.

There have been many pretty stories told about the young storks' behaviour to their parents when the old birds are weak and infirm. And it is a fact, that the young birds do succour the old ones, but whether these are their own parents remains to be proved.

At any rate, the old birds and the young live happily together until the time comes for their winter's tour. Then, a week or two before the event takes place, the storks hold a council in some field or common to discuss the matter, as we see the swallows do with us.

At length, however, the day arrives, and great flocks of storks are seen in the air flying in their peculiar manner with outstretched legs. Sometimes a flock is half a mile broad, and takes three hours in passing. They have no voice, so that the only noise they make is with their wings; and if anything

THE MARABOU STORK.

THE BOAT-BILL STORK.

startles them, they make a clattering noise with their bills that is heard at a great distance. The stork leads a very happy life, and by thus passing from one climate to another, enjoys a moderate temperature, and avoids the extremes of heat and cold.

In all countries and ages, the stork (except, indeed, in England) has received protection and kindness from man. In ancient Egypt no one was allowed to kill a stork. Its amiable and affectionate disposition have made it many friends, and it is very useful as a destroyer of reptiles. The stork, in return, fears no ill, and builds its nest amid the busy throng of men, and walks the crowded streets without the least sense of danger.

In some of the Turkish cities where the stork passes the winter, the nests are often placed on the tall round pillars of the mosques, that in Bagdad are flat at the top. Here is a famous foundation for the nest, and the pillar looks as if crowned with the nest, the head of the bird and its long neck appearing over the edge. The Turks hold the bird in the utmost esteem, and fancy that its movements resemble their own attitudes of devotion. In cities where there is a mixed population, the stork always makes its nest on the house of the Turk.

The storks spend the winter in the deserts of Africa or Arabia, and come back to their old haunts in the summer.

They build their nests on the tops of old houses or belfreys, and in the chimneys of tall houses, and even in dead trees.

In some of the marshy districts of Holland the storks are of the greatest use, for they kill the snakes and lizards and mice and frogs, that abound and are exceedingly troublesome to the inhabitants.

The grateful people encourage the stork to dwell among them, and welcome it back from its winter tour. Children sing little songs about the coming of the stork, and it is made as much fuss with as the swallow is with us.

And they sometimes make a platform at the top of the house for the stork to build her nest upon. An old cart-wheel is placed in a flat position at the top of a strong pole, and is just the very foundation the stork wants. Of course, she is only too glad to come, and very soon sticks and twigs and reeds are carried there in large quantities, for it is a very strong nest.

The outside is made of these materials, and the inside is lined with herbs, mosses, and down. Year after year the faithful and affectionate birds come to the same nest, and nothing can ever wean them from it.

THE SPOON-BILL.

THE spoon-bill has its name from the spoon-like manner in which both the upper and the lower parts of its bill terminate.

It is in other respects like the stork and the heron, and lives upon the same food. At one time the spoon-bills used to inhabit our own country, and to build their nests on the top of the tallest trees. They used to appear in March, for they are birds of passage. Many of them were destroyed for the sake, not of their flesh, but of their beautiful white plumage and their curious-shaped bill. Flocks of spoon-bills were in old times seen in the marshy land of the eastern counties, near to Yarmouth; but since much of this land has been drained and cultivated very few spoon-bills have been seen.

The birds spend their summer in Holland, and then pass into Italy or even Africa for the winter.

Their nests are made of reeds bound together by weeds, and are in the middle of the river, only a few inches above the surface of the water. The nest is not lined, and is just large enough to allow the mother bird to set on the eggs, while her partner stands beside her. Sometimes they build on high trees, and, indeed, prefer it.

They feed on fishes and insects and shrimps, and other such kind of diet; but, if pressed with hunger, will eat almost anything.

The whole of the plumage of the spoon-bill is pure white, except a band of feathers in the front of the neck that is a buff colour. It has a beautiful plume of feathers on its head. Its legs and toes and claws are black; and the toes are connected by a membrane. The beak is black, except at the rounded part, where it is yellow.

There is a curious fact about the spoon-bill that must not be passed over. It is one of the very few birds that possess no organ of voice, and it cannot utter a single note. There is an entire absence of those muscles that can contract and dilate the air-tubes by which the voice is formed and uttered; in some birds these are like a musical instrument, and enable them to pour out their songs.

THE FLAMINGO.

THERE are few birds so odd in their appearance as the flamingo.
Its body is not so large as that of the stork, but its legs are like long

THE SPOON-BILL.

stilts. Indeed, they may be said to be quite out of proportion to its size; when it stands up it is as much as six feet high. The head is small, but is furnished with a very long bill, which, as you see, curves down from the middle. The end of the bill, as far as the bend, is black, and then a reddish yellow. The tongue is large and fleshy, and fills up the whole of the bill, and the tip is gristly.

Its long legs rather link it with the waders, but the three front toes are united by a web, as in the case of the water-birds.

The plumage of the head, when in its full perfection, is deep scarlet, with black quills.

As it strides about upon its stilt-like legs, with its enormous length of neck, we should regard it as a most uncouth creature but for its splendid scarlet robe, that excites our admiration.

It lives with its companions in a flock, and the flock stand in a line, like sentinels, clad in their red uniform. One of the band acts as a watchman, and if any danger approaches, utters a scream like the sound of a trumpet. Then the whole flock rise in the air with loud clamour, and look very much like a fiery cloud.

The creeks and ravines of tropical countries in Asia and Africa abound with flamingoes. They are seen standing as in the picture, and present a most grotesque appearance. Their way of feeding is very peculiar. They twist their neck in such a way that the upper part of the bill touches the ground, while they disturb the mud with their webbed feet, and raise up the insects and spawn of which they are in search.

In the summer the flock of flamingoes will take a journey northward as far as the Rhine. When they are on the wing they have a very splendid appearance. They look like a great fiery triangle. All at once they slacken their speed, hover for a moment, and then alight on the banks of the river. They range themselves in the usual line, place their guards, and begin at once to fish.

Considering the enormous length of its legs, you would wonder how the bird contrives to hatch its eggs, or what kind of a nest it builds.

It is a mason bird, and forms its nest of mud, in the shape of a hillock, with a hole at the top. Here the mother bird lays two eggs about the size of those of the goose.

The nest is high enough to allow her to throw her legs across it and sit upon the eggs, in an attitude as if she were riding. The flamingo sitting on its nest in this manner has been compared to a man on a high stool, with his

FLAMINGOES.

legs hanging down. The nest itself is very curious, and is solid nearly to the top, and then hollow like a pot.

The bottom of the nest is in the water, and the bird usually has its feet in the water. In some parts of the tropics, the birds are tamed for the sake of their skin, which is used instead of swans' down. They are caught in snares, or else decoyed by tame flamingoes that are used on purpose. The tame flamingoes are driven into places frequented by the wild ones, and meat is laid upon the ground. As soon as the wild flamingoes see the others eating the meat, they come forward to obtain a share. A battle ensues between the birds; and the bird-catcher, who is hidden close by, watches his opportunity to dart forward and seize the prey.

There are two kinds of flamingoes—that of America is of a deep red, while the one in Asia and Africa is rosy colour, with black wings. In old times the flesh of the flamingo was considered a dainty, and even now the young bird is thought by some people to taste like partridge. But, on the whole, the people in these days, who have tasted it, say it is very oily and disagreeable.

The flamingo cannot live in England. Though many attempts have been made to rear it, it soon languishes and dies.

There was a tame flamingo that lived a little time. It used to dip its bread in water, and to eat more in the night than in the day. It was very impatient of cold, and would go so near to the fire as to burn its toes. One of its legs was hurt by an accident, and it could not use it. But it contrived to walk all the same, for it put its head to the ground and used its long neck as a crutch.

THE PEACOCK.

To see the peacock in its full beauty, he should be viewed in his own land, in the glade of some tropical forest. There he spreads out his tail of dazzling beauty, and struts about with his companions. None of the feathered tribe can vie with him in the splendour of his array. His shining colours glisten in the sun, and compare with those of the brilliant little humming-bird or the gorgeous parrot.

THE PEACOCK.

In the forests of Java the peacock wears a much longer crest than elsewhere, and there is a curious fact to be told about him. There are beautiful park-like scenes in the island, studded with great trees, and that look very smiling. But a dangerous enemy is always lurking behind some bush or fence. I mean the tiger—one of the great scourges of the island; and the traveller, as he crosses the plain, often feels his horse tremble and shake beneath him. The animal knows by instinct that the tiger is close at hand. And there are other parts of the country where villages are thinly scattered amid the wilds, and where the native lives in a state of constant dread.

The village is enclosed with strong fences, and fires are kept burning in the night. Still the enemy is not driven away. And what is very curious, the peacock and the tiger are often seen in each other's company. When the streaks of light begin to gather round, the harsh, disagreeable note of the peacock is heard, and then the villagers say that the tiger is setting out on his excursions.

The voice of the peacock is the worst part of him. It' is so harsh and discordant that there is scarcely anything like it in nature. And his disposition is not at all pleasant. He is a most destructive bird, and kills all the little chickens, and treads down all the flowers in the garden, and, unless he has a very large range, is scarcely to be tolerated. But such a range is usually provided him, and he roams about in extensive grounds and plantations, and is considered highly ornamental. At night he will roost on the highest branch of a tree, or even on the top of the house.

The peacock was first brought to us from the East Indies, and flocks of peacocks are still seen in the beautiful island of Ceylon. And its name is mentioned in the earliest history. King Solomon had peacocks brought home in his ships; and in the old days of Greece a peacock and a pen-hen were sold for as much as £30 of our money.

And during the time of Alexander the Great's expedition into India, he was dazzled by the sight of flocks of peacocks flying wild on the banks of the rivers. Their great beauty so charmed him, that he forbade any one to harm them on pain of a very severe punishment. And when the peacock was first brought to Greece, people were in raptures with it, and came from far and near to behold it.

Besides being admired for their beauty, peacocks were in those days thought a delicacy for the table. At the old Roman banquets they were the great ornaments of the feast, and served up at every entertainment. And in much later times, it was usual in France to have the bird on the table, merely as an ornament, and not to be eaten.

THE PEACOCK-PHEASANT OF ASSAM.

The skin was taken off, and the body prepared with spices. Then the skin was put on again, with all the plumage in full beauty, and in this state the peacock would keep for years. It was then placed on the table to delight the company. At wedding feasts the beak and throat of the bird were filled with cotton steeped in camphor, which was set on fire to amuse the guests. Now and then, in our own day, the peacock makes his appearance as a dainty dish on the occasion of some great banquet, but in ordinary circumstances it is never seen.

Grand as the costume of the peacock is, he is a near relative of the barndoor fowl, and feeds on grain in the same manner. But he is fickle in his tastes, and there is hardly anything he will not seek after. He eats insects, pecks off the tender buds of plants, roots up the seeds, and strips the cottage of its thatch.

In their native country whole flocks of peacocks are seen in the fields. They are very shy, and run off the minute any one comes near them. They are quicker than the partridge, and hide themselves in the thicket, where it is impossible to find them.

The fowler comes often into the thicket with a snare he has prepared to entrap the bird. He has a banner, painted with a peacock on either side and a lighted torch at the top. The peacock is frightened, and flies to what he supposes to be one of his own kind, and is caught in a noose that hangs for the purpose.

The peacocks are distinguished, as a class, by the crest on the top of the head, and the great length of the tail feathers, that can be raised and expanded, and present the most beautiful picture in nature.

THE PELICAN.

THERE are some birds, as well as animals, about whom a great many fables have been told. The pelican is one of these.

He has a great pouch under the lower part of his bill, that can hold a great many fishes. When he goes out on a fishing excursion he eats as much as he likes, and puts the rest into his pouch for the family at home.

His breast is snowy white, and the tip of the bill is red like blood; so that people have said the pelican feeds the young with his own blood

You will see that the pelican does nothing of the kind. He feeds the young ones with what he eats himself—fish.

THE PELICAN.

The pelican is a swimming bird, as you may know by looking at his feet. His four toes are webbed together like the toes of the duck or the goose. He is about the shape and colour of a swan, but much larger. The wings are

P

very large, and the bones in them very light, so that they can receive a great quantity of air, and the bird can soar very high in the air and keep a long time on the wing. The colour of the plumage is a yellowish white, and the pouch is of a straw colour. On the greater part of the head and neck the plumage is like a short close down, gradually passing into feathers; and there is a tuft on the back of the head that falls downwards over the neck.

But the pouch is the most curious part of the pelican. In the first place, the bill itself is, as you see, of an enormous length, and, if it were opened wide, a man could put his head into the bird's mouth. It is very thick in the middle, and tapers off towards the end with a little hook. From the under part of the bill hangs a bag that reaches its whole length, from the tip to the neck. When it is empty, the bird can wrinkle it up and make it go into very little room. The skin of which the bag is formed is covered with a downy substance as soft as satin. It is very elastic, and though it does not show when it is empty, it can be stretched out to almost any amount. When the pelican goes out fishing, the first thing he does is to fill his bag, and when he has done that he goes home again. He can bring as much fish home as would serve sixty hungry men for a meal!

Though the pelican is a swimming bird, he does not venture into the open sea. He lives in flocks along the sea coasts, and on the borders of lakes. He is by nature of an indolent disposition, and does not like much trouble. But hunger rouses him up from the dozing state in which he is often found, and drives him to the water-side.

When night comes, the pelicans, in spite of their webbed feet, are resolved to perch on trees to roost. They take their repose among the lighter and gayer inhabitants of the forest, and sit in a curious attitude, with the head resting on the bag, and the bag upon the breast. There they remain without changing their position, until hunger again rouses them, and they go out to seek for prey.

The mother pelican does not take much trouble about her nest, but drops her eggs on the bare ground, and hatches them there. She feeds her young with the fish in her bag that has been there some time, and is soft and tender.

A traveller once took two young pelicans and tied them by the leg to a post stuck in the ground. Here he had the pleasure of seeing the old birds come every day to feed them; and when night came they roosted on a tree that hung overhead. Indeed, both the old and young birds became so tame that the owner could both touch and handle them, and they took fish out of his hand. They always put the fish in the bag, to swallow it at their leisure.

Sometimes the pelican is made to use its pouch for the service of man. The white pelican (for there is a brown species) lives in North America, and the Indians sometimes tame and educate it. They carry its education to such a state of perfection that it goes out in the morning at the word of command, and comes back at night with its bag quite full. Its master makes it give up the plunder, and then allows it to keep part for itself.

Along the great rivers of North America the pelicans are often seen by hundreds. They will range themselves along a sand-bar, pluming their feathers, and enjoying their liberty. If one gapes, they all gape from sympathy, and open their long bills in succession in a most laughable manner.

By-and-by their toilette is complete, and their feathers are trimmed and dressed. By this time the setting sun is tinging the tall tops of the forest trees, and the birds begin to feel hungry. They rise in a clumsy way on their legs, and waddle towards the water.

Then their whole appearance changes. They are no longer clumsy and awkward, but float lightly along, propelling themselves onward with their paddle-like feet.

The pelican in its native haunts is a beautiful bird. Its eyes sparkle like diamonds, and the orange red of its legs and feet, and also of the bill and pouch, contrast with its white plumage. Its flesh is quite unfit for food.

THE CORMORANT.

WHEN the Chinaman wants a supply of fish to sell in the market, he sets out in his little boat, or junk, as it is called, down the river.

He is not going to catch the fish with a line and a hook, but in a much quicker way. And he is not going to catch them at all himself. Some old friends of his will fish for him.

His friends are a number of birds called cormorants, a name that has a sound as of some one that is very greedy. And the cormorant is greedy of fish, and will eat as much as it can swallow.

It is a large bird, with webbed feet like the pelican. But the middle toe is notched like a saw to help it hold its prey.

The feathers of the head and neck are of a jetty blackness, and the body is very thick and heavy, a little like a goose. The bill is straight till near the end, when the upper part bends into a hook. It is, as we said before, very greedy, and nothing seems to satisfy it. It would go on eating almost for ever.

But the Chinaman has a way of checking the greediness of the cormorants. He has tamed them and taught them to catch fish. But he will not trust them unless he ties a bit of string round their throats, to prevent them from swallowing what they have caught.

He goes sailing up the river in his boat, and the cormorants are all perched about him. By-and-by he stops, for he has come to a place where he knows there is plenty of fish. The cormorant sees the fish almost before his master, and down he pounces into the water. He comes up with a fish in his bill, and then his master gives him a call. The bird comes in a minute, and drops the fish in the boat. The other cormorants are busy fishing all the time, so that he soon gets his boat full. Of course, the birds expect something to eat for their trouble, and when the day's work is over they have their turn, to say nothing of a stray fish now and then that is thrown to them.

Sometimes the cormorant plays with a fish as a cat does with a mouse, letting it go and diving after it, several times, but in the end it brings it out safe enough.

The Chinese are not the only people who know the value of the cormorant as a fisher. It is said that the birds were once used in England for the same purpose.

The cormorant eats so much that it is very fat and heavy, but still it is an active bird, and is constantly flying about.

It drops down from a great height to dive after its prey, and is seldom seen except when there is fish to be had. It very seldom makes an unsuccessful dive, and is often seen rising with a fish larger than it can swallow, and sometimes it has caught the fish by the tail, which is rather awkward. In this case, the bird tosses the fish up in the air, so that it turns over, and comes head first into its mouth.

Like the pelicans, the cormorants can perch on a tree, and they are fond of sitting on the ledges of rocks, where they make their nests.

A colony of these birds nestled on the ledges of a great rocky cliff in the Gulf of St. Lawrence. A naturalist was very anxious to see them, and

THE CORMORANT.

crawled along the cliff, some hundred feet above the rolling waters, until he found himself only a few yards above the nest. He could then see the cormorants at home.

The mother fondled her young with the greatest tenderness, and put some food into the mouth of each. The little ones seemed very happy, and billed and caressed their mother.

But all at once this happy scene came to an end, for the mother bird spied the intruder, and flew away, leaving the young ones quite at his mercy. The little ones seemed much frightened, and crawled along the rock to a hole, into which they went for shelter.

The nests are made of a quantity of dry sticks, matted in a rude way with weeds and moss. The cormorants mend the old nests every year, and come back to the same place. On some of the ledges the nests are all crowded together, but on every secure place is a nest, except towards the top of the rock, where there are none.

The young birds are of a very uncouth appearance, and of a livid colour, and their legs and feet seem enormous. If any one goes near them they stretch their necks, and open their bills, and make a curious sound, between a hiss and a mutter. They crawl sluggishly about, helping themselves along by their bills, looking at all times extremely clumsy.

The cormorants perform a kind of toilet, by flapping their wings and rubbing and scratching themselves with their beak and claws, and beating up the water round themselves.

They are very regular in returning to the same place to roost, and those that have no broods spend the night apart from the rest, standing erect in files on the highest ledges.

They fly in long strings, one after the other, and are seen by hundreds in a company. It is impossible to come near them when they are fishing, they are so cautious, and have so many sentinels upon the watch.

THE SWAN.

"The swan, with arched neck
Between her white wings mantling, proudly rows
Her state with oary feet."

IN former times the swan was held in great esteem as an article of food, and used to be served up at the national feasts as the grand dish of the table. The wanton destruction of the birds, or the taking of their eggs, was severely punished in the reigns of Edward IV. and of Henry VII.

The swan possesses great strength and power in the wing. With one blow it can break a man's leg, and has been sometimes seen to dash him headlong into the water.

The gift of instinct is bestowed largely on the water birds, and enables them to escape accidents, and to overcome a great many difficulties.

THE WHISTLING SWAN.

We can tell you something about the swans that live on the banks of the Thames. At times the violent rains will cause the river to swell, and the water begins to rise. When this happens at the season for hatching, the birds have been seen busily employed in raising their nests, in order to save their eggs from being washed away by the flood.

For eighteen years a swan had built her nest by the side of the Thames, in the same spot. One spring she was sitting on her eggs as usual, when it was observed that she was getting together a quantity of grass and weeds, and trying to raise her nest.

As soon as this was noticed, a labourer was sent with a load of straw and rubbish, and told to throw it down beside her.

THE BLACK SWAN.

The bird seemed to understand what it was for, and with the materials thus provided she began at once to raise her nest some two or three feet higher.

There came a heavy fall of rain that very night, which flooded the meadows, and did a great deal of damage in the neighbourhood of the river; but the

THE BLACK-NECKED SWAN.

swan and her nest were safe. Instinct had led her to take precautions, that man, for want of foresight, had neglected. Much of his property was destroyed, but the eggs of the swan escaped, for the prudent mother had raised them just high enough to be above the flood.

The whistling swan is not quite so graceful as the tame swan, of which we have been speaking, and that is so ornamental on our lakes and rivers. The name of "whistling" is well bestowed upon it, for it sings as it flies, and the melodious notes of its song are heard at a great distance. It is a native of the colder latitudes of the North and South Seas, and during the winter months only pays a visit to this country; to the Orkney Islands and the Hebrides it comes in flocks, and great numbers are shot, and taken to market for sale.

Occasionally one has been caught alive, and will live in harmony with the ducks and geese of the farmyard.

The black-necked swan has more of the movements of the goose than of the swan. It is a native of South America, and travels to the lakes in the interior to rear its brood.

The black swan is an Australian bird, and lives in large flocks upon the rivers and lagoons of certain districts. The white man is driving it before him, and hunts it down without mercy. These birds can be kept in England, and have been known to rear their young when snow was on the ground, during the greatest cold of winter.

THE DUCK.

THE duck in a wild state is found in Europe, Asia, and America. In summer it frequents the lakes and marshes of the north, and in the winter betakes itself south to more temperate climes. In the British Isles many remain in the marshy districts all the year round.

The duck is strictly a fresh-water bird, and it is to lakes and ponds thickly covered with sedges and water plants that it generally resorts, although it is occasionally met with on the sea-coast.

Wild ducks are very gluttonous. They seem, indeed, to be always famishing with hunger, and devour everything that comes within their reach— animal and vegetable—and they are very fond of small fish. They begin to pair about March, and construct their nests in dry retired spots, under bushes or concealed herbage, near the water. The period of incubation varies from twenty to twenty-eight days, during which the female sits on the eggs with the greatest patience and self-devotion, and the ducklings are taken to the water one day after they are hatched. They grow rapidly, and in about six weeks are able to fly.

The flesh is considered very nice to eat; and a great many ways have

THE WILD DUCK.

been invented to catch the poor ducks by decoying them into a snare. Sometimes tame ducks are trained to watch for a flock, and to allure them into a net that is held open at one end a little like the tilt of a cart. The fowler is close by, and as soon as the net is full enough, he draws the end together, so as to shut it up, and catch all the ducks.

THE EIDER DUCK.

The cider duck is a native of the Arctic seas, and is found in great abundance along the shores of Iceland, Greenland, Lapland, and in Hudson's and Baffin's Bays. In the cold countries where the eider duck lives, its flesh is used for food, but we should think it had a rank and unpleasant taste.

Eider down is an article of commerce, and much in request to make into quilts and articles of clothing.

THE GREY GOOSE.

THE GOOSE.

THE goose is a bird of great value. Not only does it figure with general acceptance on the table, but its feathers and down contribute largely to the comfort and repose of many.

Although the goose is an aquatic bird, it is not exclusively so. Indeed, some geese pass the greater part of their lives on land. They walk extremely well, and though in swimming they are neither so graceful as the

THE SPUR-WINGED GOOSE.

swan, nor as active as the duck, they row themselves along the water with ease and facility, and their power of flight is remarkable.

The grey goose, from which our domestic goose is descended, was formerly common in the fenny districts of England, but is now comparatively rare both in this country and in Ireland. It is a regular winter visitor

STORMY PETRELS.

to the Orkney and Shetland Islands, but does not remain to rear its brood, passing northwards in the spring.

Geese live principally on vegetable diet; by means of the hard margins of their beaks they are able to crop grass and various kinds of vegetables. They likewise procure small animals and plants from the bottom of the pond or river.

The spur-winged goose, of which we give an illustration, is a native of Central and South Africa. It is very rarely to be seen in England, although specimens have been shot in Yorkshire, and also in Cornwall.

THE STORMY PETREL.

THE stormy petrel is a bird of the sea, and very different in its habits from the birds we have been describing. In colour it is of a sooty black. Its wings are like those of the swallow, and enable it to mount high in the air; and they are so strong that it can keep all day on the wing, and may be heard throughout the night. Indeed, it rarely comes to land, except to make its nest in some cliff or steep rock at a great height above the sea. And it can hardly be said to swim. The fore toes are webbed, and the hinder toe is a mere claw. It runs about upon the surface of the waves, or sits down and floats along.

The sailors call the petrel by the name of "Mother Carey's chicken," and dread to see one near their ship. They say the bird foretells a storm, and is come to wait for the wreck that it may feast upon the prey. But the sailors are mistaken. In time of danger the petrel is seen in the track of the vessel, but it is only to obtain a shelter, by placing the ship between itself and the tempest.

In disposition the petrels are very harmless. They live in perfect good-fellowship with each other, and rarely seek the company of other birds. Their food consists of all sorts of soft-bodied animals, picked up from the surface of the ocean.

www.ingramcontent.com/pod-product-compliance
Lightning Source LLC
Chambersburg PA
CBHW021409230426
43666CB00006B/691